THE KNOWLEDGE WE HAVE LOST IN INFORMATION

THE KNOWLEDGE WE HAVE LOST IN INFORMATION

The History of Information in Modern Economics

Philip Mirowski

AND

Edward Nik-Khah

OXFORD
UNIVERSITY PRESS

OXFORD
UNIVERSITY PRESS

Oxford University Press is a department of the University of Oxford. It furthers the University's objective of excellence in research, scholarship, and education by publishing worldwide. Oxford is a registered trade mark of Oxford University Press in the UK and certain other countries.

Published in the United States of America by Oxford University Press
198 Madison Avenue, New York, NY 10016, United States of America.

© Philip Mirowski and Edward Nik-Khah 2017

All rights reserved. No part of this publication may be reproduced, stored in a retrieval system, or transmitted, in any form or by any means, without the prior permission in writing of Oxford University Press, or as expressly permitted by law, by license, or under terms agreed with the appropriate reproduction rights organization. Inquiries concerning reproduction outside the scope of the above should be sent to the Rights Department, Oxford University Press, at the address above.

You must not circulate this work in any other form
and you must impose this same condition on any acquirer.

Library of Congress Cataloging-in-Publication Data
Names: Mirowski, Philip, 1951- author. | Nik-Khah, Edward M., author.
Title: The knowledge we have lost in information: the history of information in modern economics / Philip Mirowski & Edward Nik-Khah.
Description: New York City: Oxford University Press, 2017. | Includes index.
Identifiers: LCCN 2016036012 | ISBN 9780190270056 (hardcover)
Subjects: LCSH: Neoliberalism. | Neoclassical school of economics. | Practical reason. | Economics.
Classification: LCC JC574.M567 2017 | DDC 338.501—dc23 LC record available at https://lccn.loc.gov/2016036012

1 3 5 7 9 8 6 4 2
Printed by Sheridan Books, Inc., United States of America

CONTENTS

List of Figures and Tables — vii
Acknowledgments — ix

1. It's Not Rational — 1
2. The Standard Narrative and the Bigger Picture — 31
3. Natural Science Inspirations — 45
4. The Nobels and the Neoliberals — 51
5. The Socialist Calculation Controversy as the Starting Point of the Economics of Information — 60
6. Hayek Changes His Mind — 66
7. The Neoclassical Economics of Information Was Incubated at Cowles — 73
8. Three Different Modalities of Information in Neoclassical Theory — 101

CONTENTS

9. Going the Market One Better — 124
10. The History of Markets and the Theory of Market Design — 144
11. The Walrasian School of Design — 161
12. The Bayes-Nash School of Design — 170
13. The Experimentalist School of Design — 183
14. Hayek and the Schools of Design — 193
15. Designs on the Market: The FCC Spectrum Auctions — 207
16. Private Intellectuals and Public Perplexity: The TARP — 221
17. Artificial Ignorance — 233

Notes — 243
Bibliography — 271
Index — 293

LIST OF FIGURES AND TABLES

FIGURES

3.1	Shannon Information Theory	46
8.1	The Block Universe of Relativistic Physics	110
8.2	The Harsanyi Setup	114
10.1	Orthodox Trajectory Through Information Space, I	153
10.2	Orthodox Trajectory Through Information Space, II	159
11.1	Reiter's Schematic	165
12.1	An Equilibrium Bidding Strategy, According to the Bayes-Nash Approach	177
14.1	Orthodox Trajectory Through Information Space, III	204

TABLES

4.1	Economics of Information/Knowledge Nobel List, 1969–1994	52
7.1	Cowles Commission Members and their Information Enthusiasms	80
8.1	Three Formal Approaches to Information	102
15.1	Auctionomics' Product Comparisons	220

ACKNOWLEDGMENTS

We were inspired to write this book after accepting an invitation extended by the Institute for New Economic Thinking to present a short course on the history of twentieth-century economics. Both of us, jointly and separately, had written on themes relevant to this topic: the confusion evident in the sundry treatments of information by economists, the neoliberalization of the economics profession, the constructivist turn in economics, and the largely misunderstood legacy of experimental economics, among others. Our intention was to summarize this previous work and draw from it lessons relevant to our post-crisis world: without understanding where orthodox economics is headed, how can the student get excited about the "new"?

Thus, we set out to explain how the economics profession ended up in its current state, seen from a doctrinal perspective. We already had bits and pieces of the story in hand, but setting out to weave them into a coherent and (we hope) compelling narrative required that we push far beyond anything we had previously written. What you hold in your hands is the result.

ACKNOWLEDGMENTS

We presented an early draft of this book at the Hong Kong meetings of the Institute for New Economic Thinking (INET) and a revised version at INET's New York studios. We thank INET for supporting those lectures: this book benefited from the highly engaged audiences at these two events. Edward Nik-Khah is grateful to the Roanoke College Faculty Research Year program for support to complete this manuscript. Also, we thank Kyu Sang Lee for providing helpful suggestions at a critical juncture. Both of us wish to thank the Duke Center for the History of Political Economy for a resident scholar fellowship, which permitted us to work through some of these arguments at length in person.

We also want to acknowledge the encouragement of Scott Parris, now retired, one of the most important academic editors of the history of economics during his tenure at Cambridge University Press and then Oxford University Press. We will miss you, Scott.

This book makes use of passages from a handful of previously published articles. Chapter 15 draws from "A Tale of Two Auctions," *Journal of Institutional Economics* 4(1): 73–97. Chapter 16 includes excerpts from "Private Intellectuals and Public Perplexity: The Economics Profession and the Economic Crisis," which appeared in *History of Political Economy*, Vol. 45 (suppl 1): 279–311.

[1]
IT'S NOT RATIONAL

> The most merciful thing in the world, I think, is the inability of the human mind to correlate all its contents. We live on a placid island of ignorance in the midst of black seas of infinity, and it was not meant that we should voyage far. The sciences, each straining in its own direction, have hitherto harmed us little; but some day the piecing together of dissociated knowledge will open up such terrifying vistas of reality, and of our frightful position therein, that we shall either go mad from the revelation or flee from the deadly light into the peace and safety of a new dark age.
>
> H. P. Lovecraft, "The Call of Cthulhu," 1928

This book is mostly concerned with the history of economics; but we would like to suggest at the outset that it also describes a cultural rupture of far larger import. To a first approximation, it explores how economists changed what it meant within their discipline to claim to "know something," and consequently to lay claim to a special kind of expertise at the dawn of the twenty-first century. But this did not happen in a vacuum. Not to sugar-coat what might be a somewhat unpalatable assertion, what it meant to "know the truth" changed dramatically and irreversibly after World War II. In saying this, we are not engaging in the usual hand-wringing concerning postmodernism and cultural relativism that pundits have bewailed from the 1990s onwards. After all, distancing oneself from the truth claims made by historical protagonists is just something all good historians do; there is nothing that especially is distinctive or scandalous about agnosticism in the modern era. Rather, our concern in this book is with the postwar changes in the perceived validation of the truth mediated by the rise of "information" in the social sciences, and especially in economics. The truth, as conceived by modern economists, has not set anyone free. Instead, it brought about the death of the Kantian subject, and a subsequent lifeworld hollowed out the humanist concerns that many people mistakenly think are the heart and soul of a science of economics.

WHAT IS TRUTH IN ECONOMICS?

Loose talk about "truth" is bound to make most people, and many economists, skittish in the extreme. Talk about "information," by contrast, would seem far less threatening; and rest assured, most of this book will be couched in the more soothing idiom of "information" because that is how our protagonists preferred it. But we would be shirking our duty to the reader if we did not admit that just beneath the surface of our narrative lurks the suspicion that the surfeit of talk about information serves to obscure something more essential, which for purposes of this introduction we will intermittently call "knowledge," or more brutally, Truth. Given that the history which follows will present us with the most variegated conceptions of what it means to "know" something in economics, a few preliminary observations about our own philosophical position might be in order.

The postwar worry about truth in economics kicks off with a relatively famous 1940 article by Chicago economist Frank Knight entitled, appropriately, "What Is Truth in Economics?" Knight wanted to get his peers to think a bit harder about what they rather cavalierly would endorse as Truth.[1] One major concern for Knight was the suspicion that "liberalism" might suffer from debilitating internal contradictions, such as the incompatibility of the search for Truth and commitment to a freedom for anyone to think what he or she likes. Back then, Knight was fighting a losing rear-guard battle against the rising tide of logical positivists of the era; he feared a situation where "truth is merely a game in which the players are free to make any rules they please."[2] Reading his paper now, it beggars the imagination that anything nearly so philosophically self-critical could ever be published in the *Journal of Political Economy* these days. One reason for this reversal is that modern economists appear no longer capable of hard thinking about the nature of Truth; the best they can manage at Chicago, it seems, is to argue that "good economists" enjoy a high degree of consensus about economic matters when responding to questionnaires, so not to worry.

To illustrate this, we seek to briefly contrast the bygone Chicago of Frank Knight with the contemporary Chicago of Luigi Zingales and collaborators, who in 2013 published an article comparing the responses of forty-one faculty at a very few high-ranked U.S. research universities, with a sample of U.S. households conducted by the Chicago Booth Financial Trust Index project. Their headline was that there subsisted remarkable consensus among their sample of economists, but not with their sample of the lay public. Zingales and co-author Paola Sapienza reported a striking thirty-five percentage point gap, on average, between the economists' beliefs and the public. For example, about three out of four of the general public respondents said that a "Buy American" policy is good for manufacturing employment, while only 11 percent of economic experts agree. Nearly all the economists queried avowed that the North American Free Trade Agreement (NAFTA) has helped Americans prosper, but only half of respondents to the Booth survey thought so. Zingales seemed loathe to admit that the credibility of the economics profession may have suffered somewhat following the global economic crisis, still fresh in the minds of his interlocutors, and this might account for the gap.

This paper in a modern idiom is wildly popular in the economics literature these days, because it reinforces the orthodox self-image of the economist. First off, there is the extraordinary conviction that all "real economists of sound instincts" essentially agree on everything, when in fact what actually happens is that boundaries of orthodoxy are continuously being policed by a few economists located at a few top-ranked departments; it follows that their hand-picked peers are effectively self-selected for consilience; and thus appeals to consensus turn out to be effectively tautologies. Even Frank Knight knew that reliance on consensus was the lazy man's definition of Truth. But second, there lurks a barely repressed contempt for the beliefs and opinions of the general public. Once upon a time, it was permissible to presume economic agents were pretty smart, and therefore of sound mind; but no longer. This curious about-face within the modern economics profession is one of the major themes of the present volume. The mid-century Walrasian orthodoxy came clad with all sorts of "welfare theorems" that insisted markets always and everywhere gave the people what they wanted; but as the "information" revolution began to suggest that market participants didn't really know very well what they wanted, then for the first time in history, economists began to assert their competence to "design" markets, with the objective of giving people what economists believed they *should* want.

This turns out to be something of far greater import than some passing dalliance with mere abstract epistemology: as it happened, it underpins the very politics of the modern profession. One explanation for economists' recurrent tendency not to trust democracy, for instance, is that they suspect the man in the street is an epistemic shambles; in their estimation, economists therefore deserve to be respected as experts in knowledge, because their training encourages them to approach the reasoning of the layperson with a cold jaundiced eye. By contrast with economists' perception of their own situation of purported unanimity, any consensus they happen to find among the unwashed is no index of anything whatsoever.

Modern economists love this self-portrayal of their blessed status of epistemic expertise, but it is false in every respect. All you have to do is read the newspapers to realize individual economists have been persistently at each other's throats; the recent crisis merely brought this situation closer to consciousness for the public.[3] If there is widespread adherence to some doctrines among economists, that fealty tends to be more in the nature of ceremonial obeisance than carefully considered conviction.

Let us point to just one example, to prepare the ground for our history. All economists believe in the "laws of supply and demand," right? Every parrot and TV reporter blandly repeats it as gospel truth. But those who have some appreciation for the history of economic theory, and especially regarding the Sonnenschein/Mantel/Debreu theorems, which you can find in many graduate microeconomic textbooks, are also aware that those theorems essentially obviate the existence of any single valued smooth demand curve. We do not aim to provide a history of the SMD theorems in this book, because it wanders a bit far from our mandate.[4] All we want to suggest here is that neoclassical economists have been known to subscribe to contradictory propositions that demand curves both do and do not exist, simultaneously. Epistemic flexibility goes with the territory.

There is a hermeneutic attitude that will prove conducive to apprehension of our history: the precept we are suggesting is that economists nowadays possess a rather louche attitude toward truth. We do not approach this current history as an occasion for rabid "gotcha" exercises against the veridically challenged; rather, we want to ask: What sort of profession treats payments on the side and conflicts of interest as essentially harmless, as Gerry Epstein and George DeMartino have documented, and considers a code of ethics as something only other lesser mortals may need? What sort of person denies economics is an agonistic field? What kind of orthodoxy seems comfortable with characterizing the human subjects of their prognostications as "mindless"?[5] What can it imply when a recent winner of the prestigious John Bates Clark Medal writes, "in the context of a persuasion game, so long as there is one provider of information in every state of nature that would prefer for consumers to have accurate beliefs, the truth will always be revealed to a consumer to access with reports from all providers"?[6] What sort of intellectual revels in the notion that he will never suffer anything more than fleeting transient embarrassment (because the public has a notoriously short memory) for statements of dubious veracity, confident no one will ever fire him for incompetence from a central bank, nor shut down his university economics department as a cost-saving measure, nor force him to run the gauntlet of a public shaming exercise? Or, with more direct reference to the topic at hand in this volume, what can it mean for some economist comfortably ensconced at the Institute for Advanced Study in Princeton to write, "Ideas are strangely absent from modern models of political economy."[7] In other words, we aim to echo Frank Knight's original query: What is truth in modern economics?

The answer deserves something approaching the measured philosophical and self-critical consideration of a Knight, something which does not sit well in contemporary discussions of economics. While we shall not engage in much in the way of explicit philosophizing in this volume, we shall describe in subsequent chapters how cultural trends and scientific developments helped usher the economics profession from a period right after World War II—when there was great uncertainty about what, if anything, they were capable of saying with confidence about information, knowledge, and truth—into the modern situation, which has converged on a very peculiar set of epistemic doctrines. The main task of this volume is to explain how we got from there to here.

Let us oversimplify, in the interests of inviting the reader to sample our subsequent chapters. For orthodox economists today, truth is not a matter of morality, nor of individual standards of veracity, nor even coherence with some simplistic notion of the scientific method. For the orthodox economist, core doctrine dictates that truth is the output of the greatest information processor known to humankind—namely, The Market. From the efficient markets hypothesis to Nash equilibrium in game theory, to rational expectations macroeconomics to the multiple schools of market design, the twenty-first-century economist testifies over and over again that it is The Market alone that effectively winnows and validates the truth from a glut of information. The hapless agent may or may not have ambitious epistemic pretensions; so-called behavioral economics preaches that the agent is beset with biases and lapses of attention; but the wise market participant always defers to the pronouncements of the market. Paraphrasing economist and Mont Pèlerin Society[8] member Robert Barro, as long as they keep paying us, we must be right.[9] Pelf makes right, not might.

Yet it is the next step in the syllogism that has turned out to be truly novel. If markets indeed validate truth, then the cadre that gets to construct the markets gets the final say on the nature of truth. The visible hand that fashions the auction believes it can govern the world.

TALES OF RATIOCINATION

It will probably come as no surprise that we personally do not accept the economist's imprimatur of The Market as the final solution to the age-old problem of "What is Truth?" Thus do we owe the reader some brief cursory indications of the alternative stance toward truth that governs our principles of selection in this history. Contrary to academic expectations, it may be helpful to note we do not fall back on the Philosophy 101 version of "justified true belief" as the bedrock for our various narrative choices in this history of "information."[10] It strikes us that the pertinent organizing principles are not timeless monolithic criteria such as those often championed in Philosophy 101 but, rather, they involve acknowledgment that epistemology has meant different things to different groups in intellectual history.

Perhaps the type of philosophical rupture we have in mind bears a family resemblance to the notion of *parrhēsia*, the topic of Michel Foucault's last lectures.[11] He defined the term as the analysis of practices of telling the truth about oneself; what makes that intersect with our current concerns is that he also proposes that the notion of *parrhēsia* was "originally rooted in political practice and the problematization of democracy, then later diverging towards the sphere of personal ethics and the formation of the moral subject" (2011, p. 8). While Foucault certainly did not entertain any parallel equivalent modern rupture back in 1984—there is only so much prescience one can attribute to Foucault, even given his well-known foresight concerning neoliberalism—here we are intent upon stressing the inescapable connection of politics and skepticism about democracy in which the modern transformation of truth is deeply rooted. In short, whereas Foucault was mostly intent upon comparing Greek thought to the later Christian and Cynical developments concerning care of the self, we are instead fascinated with the ways a seemingly technical neutral notion like "information" has been slowly changing what it means to "know something" and by the twenty-first century has undermined liberal secular notions of democracy and Kantian notions of the ethical self. Hence, we are open to the possibility the history we proffer here shares certain Foucauvian ambitions with regard to "genealogies": to clarify how that might work, let us dally briefly with the genre of detective stories.

The metamorphosis of the detective/spy in modern literature is not often something the average economist takes time out to contemplate. [12] A little reflection would nevertheless reveal that the "classical" detective tended to be portrayed as a super-intelligent (if a bit quirky) soul who would pick up on the little clues everyone else—and especially the plodding copper—would overlook. From Conan Doyle's Sherlock Holmes to John Buchan's Richard Hannay in the twentieth century, it was the burden of the superior individual to piece together the shards of history so as to arrive at the truth concerning guilt or innocence. The same went for spies, from Dashiell Hammett's Continental Op to Ian Fleming's James Bond. The reader went along for the ride, with the game being to see if you could outguess the gumshoe or spook as to whodunit before the story came to its conclusion. But the superhuman feats of ratiocination began to lose their luster by the middle of the twentieth century, to be replaced by a different sort of spy narrative.

As Rob Horning (2012) reminds us, a curiously different sort of spy popped up in literature around that time. He cites the work of Eric Ambler—*Epitaph for a Spy* (1938), *Cause for Alarm* (1938),

Journey into Fear (1940)—as a harbinger of this trend. To quote Horning:

> All of them feature ordinary, slightly disreputable men who more or less inadvertently end up in the middle of international security conspiracies, accused of crimes they hadn't known they committed, fleeing corrupt and/or incompetent police, or working in coordination with other foreign agents whose trustworthiness remains undecidable.... The 1930s brought the kind of war in which every member of society was indiscriminately targeted for death from above. This would provoke a climate of militant prudence and ambient mistrust in which, say, British citizens were expected to destroy any household maps and falsify local signage to confuse expected invaders.

Ambler's novels reflect this growing anxiety over protecting information, brought on both by technological developments that made it easier to disseminate information and by the entangled complexity that dispersed relevant data across a broader populace. In *Epitaph for a Spy*, the protagonist's mere possession of a camera embroils him in an intelligence investigation and he is forced to scheme how to out a foreign agent. *Cause for Alarm* centers on a machine-company sales rep who finds himself with access to sensitive armament data. Graham, the hero of *Journey into Fear*, is targeted for assassination because his engineering work makes him know too much when war breaks out. Ambler's protagonists rarely know that they know something important; the news is generally broken to them through a violent attack or an arrest. They then learn they have become intelligence agents against their will—they have become the unwitting conduit of vital knowledge that can be transmitted through them without their being capable of understanding its broader importance.[13]

"Intelligence agents against their will"—what an apt turn of phrase! Instead of possessing some special transcendental capacity to discern the truth which renders them an ideal spy, the new model protagonist is a schlub who comes equipped with little more than mediocre intelligence, but is nonetheless thrust into a whirlwind of deception and secrets. The basic plot point is intended to induce vertigo: you, the protagonist, have no idea what you are doing, but no one but you are able to do this. The leading man's meager moiety of information seems insignificant, but opens a crack to view an unseen world, such that he is caught up in forces beyond his ken which render that information (and therefore his life) so critical that the protagonist must risk everything. The meaning and significance of his appointed task may not always become fully apparent to the bumbling protagonist, but shadowy players and obscure forces recruit them as unwitting conduits for history.

Michael Chabon nicely summarizes the standard plot skeleton:

> At first the problem sounds manageable. The sleuth agrees to look into it, make a call, drop in on someone. In the end, after many neighborhoods and social strata (always coextensive in a private-eye novel) have been traversed and visited, and after a vivid array of toughs, losers, and the occasional innocent has been plotted along intersecting axes of power, money, and lust, the original problem turns out to go much deeper, and much higher, than the sleuth or the reader reckoned. That original problem was only a loose thread, it turns out, and when the sleuth tugs on it the world unravels.[14]

The place and perspective of the reader are clearly different with the onset of this new genre. In most cases, the question of whodunit is not really all that important; indeed, that might be revealed at any juncture in the narrative. Anyway, there is no great thrill in outguessing a stumblebum. Some key facts might come to light in the course of events; but equally, cabals and connivance are rarely wrapped up in a tidy package at the end. The thrill for the reader seems to come in imagining being caught up in something of world-historical significance that he or she had previously never suspected: you, too, could become a "secret agent" by being in the wrong place at the right time. Suddenly, any nondescript bush-leaguer can make a difference. The mediocre cog is elevated by Providence, or maybe just the hand of history. We all avowedly profess to believe in the agency of the individual, which would imply that we judge personal choices with respect to outcomes; but the truth of the late modern detective novel is that stark and simple causal chains are denied to most of us. Insights of lasting consequence come out of left field, unheralded and unbidden.[15]

What makes Horning's thesis so striking is that he notices two things about the rupture in spy novels that turn out to be absolutely central to the history of economics we recount herein: one has to do with the technical aspects of information, and the other with politics. We quote Horning on the first point:

> The spylike pursuit of information rather than knowledge makes us function less as thinkers than processors, personal computers—and inefficient, low-powered ones at that. We are not the subjects who know things or intentionally produce knowledge; we are instead means of circulation—objects through which information passes with more or less noise in the signal. We become not only part of a network but part of a circuit. We are pawns in a larger game, "a fly caught in the

cog-wheels" as Vandassy, the narrator of *Epitaph for a Spy*, puts it.[16]

This information, this elusive something which we somehow possess while not quite understanding it, has indeed become a hallmark of the modern predicament. Rather than believing that the "truth shall set us free," we now suspect rather that the truth, if it be such, keeps us in our place. Since we agents are no longer expected to be able to comprehensively validate information, or recognize its worth, it takes on an aura of existence independent of what we think about it. With some nudging from the computer, this has been made manifest in the contemporary phenomenon of an alienated information—something that takes on a life of its own, a hypostasized entity that has its own dimensions and metrics. The best we can hope for is to sneak up on it, like a spy, and catch it in *flagrante delicto*.

And then there is the political point. Horning makes the astute observation that this inversion of the spy story did not come out of the clear blue, but tracked an important change in political theory. He perceptively cites the work of Friedrich Hayek, who *at the very same time* was describing an economic protagonist who only possessed partial and incomplete knowledge of the economy but was co-opted into the larger conspiracy of The Market to pursue ends about which he was only vaguely aware. People were not blazingly rational, said Hayek, but they possessed limited cognitive abilities. Information was being shuttled hither and yon behind the backs of traders; they only glimpsed the flash and gleam out of the corners of their eyes. Government was just another of the shadowy forces pushing the dim individual from pillar to post; the argument against the cold war enemy was that he would not acquiesce to the ineffable wisdom of The Market; and infected with hubris, he could never know that he was badly mistaken. Big organizations everywhere were lurching around in the dark; people risked becoming cogs in an inhuman machine. "Agents" in orthodox models of economics were thus being repurposed as spies in the House of Gov; someday a real revolution would eject all those misguided souls from government who believed they could control the tides of history, or so said Hayek.

Once we observe how human agency became diminished in the modern spy novel, as information becomes reified and hypostasized, it comes as a shock to realize the same thing has happened in neoliberal political theory, and then, with a lag, also in economics.[17] Economic agents were getting lost in the Big Forces that swirled all about them. Democracy was no longer considered the bulwark of progress in both instances, because the little guy might not be depended upon to do the right thing in dire circumstances. Governments were portrayed as risible attempts to control the ever-ramifying conspiracies of citizens; faceless bureaucrats never were capable of understanding the real meaning of events until it was too late. Only The Market knew for sure. And what it knew was "information."

TINKER, TAILOR, SOLDIER, ECONOMIST

The history of economic thought often finds itself nostalgic for the older spy genres, as though the culture had never moved on. If the economic agent might seem to have become a little addled, in the orthodox frame-tale the neoclassical economist never succumbs to similar disorientation. An older, and still very popular, mode of recounting the saga of economics is constructed around hagiographic tributes to inscrutable geniuses, who see their way to truths denied to others, largely by dint of their own exquisite perception and superior intelligence. They are the Sherlock Holmeses of social thought, the detectives of pecuniary life, making connections in a manner that runs rings around the plodding proponents of pre-modern economics, not to mention the other social sciences. Deductions lead cleanly from one to another, in lockstep. However popular for ceremonial public purposes (like Bank of Sweden award lectures), these narratives exude a fusty outmoded air and stifle narrative drive with complacency.

This is not the way we opt to tell the story of modern economics. Instead, we endorse the newer breed of intelligence agents as *dramatis personae*, and consequently approach the protagonists in their often clueless states, touched by forces beyond their ken, recruited to be undercover proponents of a New World of information in economics. We believe the rise of information as an organizing principle for understanding the economy and politics was first and foremost a *cultural* phenomenon, stretching from the natural sciences to economics to, yes, spy stories. Economists could no more evade the tendencies that swept them along than they could declare themselves independent of the stochastic worldview or the triumph of abstraction in the arts. But this would imply that the history of economics was not solely or even primarily the working through of logical implications of some abstract mother-structures of economic life, such as, say, the Arrow-Debreu model of general equilibrium, or the Euler equations of intertemporal optimization. As information swept through the discipline, economists could not altogether escape the cognitive challenges that they were blithely projecting onto their models of agency.

Those who seek to reinforce the older-style histories have struggled to come up with adequate categories to encompass the blooming, buzzing confusion over the profusion of exercises that call themselves the economics of information and/or knowledge. One recent example, by Samuli Leppälä (2015), seeks to divide the theoretical endeavors into those concerned with "technological knowledge" and those concerned with "market knowledge." We don't really think the distinction is historically or logically tenable, although we do have some idea of what he means in attempting this. It has certainly been the case that there has grown up a large literature concerned with something called "technological change," which of necessity occurs at a more macro level, bound up with abstract production functions and growth theory. Another separate, but massive literature tends to approach questions of knowledge and information at a more "individual" level, often traveling under such rubrics as "decision theory" or the "economics of information," and is more explicitly tied to neoclassical microeconomics. The trouble with treating them separately is that broad conceptions of the analytical character of information have tended to move in tandem through both areas during the postwar era—and that evolution is the story we tell in this volume. Nevertheless, it is certainly true that the former class of theories often become tangled up in images of what Science really is, and the putative lessons of Nature for grounding the Economy. Since one of us has covered this particular history in some previous work,[18] we shall in this volume tend to give concepts of "technological change" short shrift, in favor of questions of epistemology in microeconomics.

The reader will thus encounter herein a very different sort of history than has been conventionally on offer when it comes to contemplation of economics in retrospect. Not only will we avoid the usual reduction of economic thought to personal genius and its travail, but we shall also invert the usual strategies of writing the history of modern microeconomics. For decades now, it seems almost obligatory that, once students have learned a smattering of neoclassical price theory, they become convinced of the banality that microeconomics is really about the formal consequences of "rationality." More often than not, this leads to interminable arguments over whether people are really "rational" or not, something concerning which apparently everyone feels fully qualified and capable of having an opinion. Rather than engage in caricature, let us sample an exemplary instance, hot off the blogs:

> Neoliberalism died in 2008 with the cratering of the global economy. With support from their sponsors at the Royal Society of Arts' 'Social Brain Unit', the signal was loud and clear: Today, new advances in 'behavioural economics' and 'neuroeconomics' drawing on the 'interdisciplinary' 'pluralistic' insights of evolutionary psychology, bio-anthropology and cognitive science point the way to the future.
>
> Forget those idiotic economists who think everyone is rational! Haven't you ever seen a TED talk? Harvard psychologist Steven Pinker says we have ended violence thanks to landmark discoveries which point to our 'hardwired' irrationalities! The future* of economics and public policy lies in the discovery of these biological characteristics implanted when our ancestors were running away from a sabre-toothed tiger on the African savannah.

Neoliberalism and public choice? Forget about it. Who even needs choice when there's no rationality?[19]

This impression of the "great liberation" of neoclassical orthodoxy from *Homo economicus* is one of the stranger consequences of the triumph of "information" in modern economics. Ominously, similar sorts of sentiments govern much modern historical work on the intellectual lineage of twentieth-century microeconomics, albeit at a much higher level of sophistication. Back when we set out to research the history of information in economics, most of what we encountered were texts that sought to trace the "history of rationality" instead. Some of the best of a rather uneven bunch are *Modeling Rational Agents* (Giocoli 2003), "Producing Reason" (Heyck 2012), *How Reason Almost Lost its Mind* (Erickson et al. 2013), and *Behavioral Economics: A History* (Heukelom 2014). Each chronicle argues that the key to understanding modern histories of neoclassical microeconomics, or alternatively, "decision theory," was to distill the massive scholarly archive down to the humanist question of what was thought, in sequential eras, it meant for a person to be a rational human being. Almost invariably, the historian in question disparages mechanistic portrayals of human rationality dating from the nineteenth century, and rejoices in the superior enlightenment of the present, wherein the economics profession has finally come to appreciate that humanity is far more antipodean and paradoxical, richly emotional and multirational than previously thought. While we acknowledge that some researchers have prided themselves on pushing the boundaries of human rationality, they mostly misunderstand their own role in the larger dynamic of the intellectual history of economics. The inversion of these upbeat narratives that we put forward in this volume will entertain the proposition that *human rationality has become increasingly irrelevant to the content of microeconomics*, and that much of this trend has been rendered plausible through the instrumentality of reconceptualizing markets as information processors. In such a context, "behavioral economics" turns out to be a sideshow.

Given the massive literature on so-called rationality in the social sciences, it gives one pause to observe what a dark palimpsest the annals of rational choice has become. The modern economist, who avoids philosophy and psychology as the couch potato avoids the gym, has almost no appreciation for the rich archive of paradoxes of rationality. This has come to pass primarily by insisting upon a distinctly peculiar template as the necessary starting point of all discussion, at least from the 1950s onwards. Neoclassical economists frequently characterize their schema as comprising three components: (a) a consistent well-behaved preference ordering reflecting the mindset of some individual; (b) the axiomatic method employed to describe mental manipulations of (a) as comprising the definition of "rational choice"; and (c) reduction of all social phenomena to be attributed to the activities of individual agents applying (b) to (a). These three components may be referred to in shorthand as: "utility" functions, formal axiomatic definitions (including maximization provisions and consistency restrictions), and some species of methodological individualism.

The immediate response is to marvel at how anyone could have confused this extraordinary contraption with the lush forest of human rationality, however loosely defined. Start with component (a). The preexistence of an inviolate preference order rules out of bounds most phenomena of learning, as well as the simplest and most commonplace of human experiences—that feeling of changing one's mind. The obstacles that this doctrine pose for problems of the treatment of information turns out to be central to our historical account. People have been frequently known to make personally "inconsistent" evaluations of events both observed and unobserved; yet in rational choice theory, committing such a solecism is the only real mortal sin—one that gets you harshly punished at minimum and summarily drummed out of the realm of the rational in the final analysis. Now, let's contemplate component (b). That dogma insists the best way to enshrine rationality is by mimicking a formal axiomatic system—as if that were some sterling bulwark against human frailty and oblique hidden flaws of hubris. One would have thought Gödel's Theorem might have chilled the enthusiasm for this format, but curiously, the opposite happened instead. Every rational man within this tradition is therefore presupposed to conform to his own impregnable axiom system—something that comes pre-loaded, like Microsoft on a laptop. This cod-Bourbakism[20] ruled out many further phenomena that one might otherwise innocently call "rational": an experimental or pragmatic stance toward the world; a life where one understands prudence as behaving different ways (meaning different "rationalities") in different contexts; a self-conception predicated on the possibility that much personal knowledge is embodied, tacit, inarticulate, and heavily emotion driven. Furthermore, it strangely banishes many computational approaches to cognition: for instance, it simply elides the fact that much algorithmic inference can be shown to be non-computable in practice; or a somewhat less daunting proposition, that it is intractable in terms of the time and resources required to carry it out. The "information revolution" in economics primarily consisted of the development of Rube Goldberg–type contraptions to nominally get around these implications. Finally, contemplate component (c): complaints about methodological individualism are so drearily commonplace in history that it would be tedious to reproduce them here. Suffice it to say that (c) simply denies the very existence of social cognition in its many manifestations as deserving of the honorific "rational."

There is nothing new about any of these observations. Veblen's famous quote summed them up more than a century ago: "The hedonistic conception of man is that of a lightning calculator of pleasures and pains, who oscillates like a homogeneous globule of desire of happiness under the impulse of stimuli that shift him about the area, but leave him intact."[21] The roster of latter-day dissenters is equally illustrious, from Herbert Simon to Amartya Sen to Gerd Gigerenzer, if none perhaps is quite up to his snuff in stylish prose or withering skepticism. It is commonplace to note just how ineffectual their dissent has been in changing modern economic practice.

Why anyone would come to mistake this virtual system of billiard balls careening across the baize as capturing the white-hot conviction of rationality in human life is a question worthy of a few years of hard work by competent intellectual historians; but that does not seem to be what we have been bequeathed. In its place sits the work of (mostly) historians of economics and a few historians of science treating these three components of rationality as if they were more or less patently obvious, while scouring over fine points of dispute concerning the formalisms involved, and in particular, an inordinate fascination for rival treatments of probability theory within that framework. We get histories of ordinal versus cardinal utility, game theory, "behavioral" peccadillos, preferences versus "capacities," social choice theory, experimental interventions, causal versus evidential decision theory, formalized management theory, and so forth, all situated within a larger framework of the inexorable rise of neoclassical economics. Historians treat components (a–c) as if they were the obvious touchstone of any further research, the alpha and omega of what it means to be "rational." Everything that comes after this is just a working out of details or a cleaning up of minor glitches. If and when this "rational choice" complex is observed taking root within political science, sociology, biology, or some precincts of psychology, it is often treated as though it had "migrated" intact from the economists' citadel. If that option is declined, then instead it is intimated that "science" and the "mathematical tools" made the figures in question revert to certain stereotypic caricatures of rationality.[22]

Beyond that, there is the even more vexing phenomenon that this abstruse definition of "rationality" is simply taken for granted as prelude for making further generalizations about the trajectory of economics after 1980, most of which suggests that the current generation of economists has providentially become the most open-minded, subtle, and psychologically sophisticated researchers in all the annals of economic thought, now that they have managed to perform the astounding conjuring trick of somehow augmenting and reconciling the previous mechanical construct of "rationality" with all manner of behavioral quirks, sociological idiosyncrasies, mental peccadillos, and outright "irrationalities" that beset the cognitive makeup of the jittery *Homo economicus*. With the help of some hefty hardware, such as PET scanners and magnetic resonance imagers, they have the brain in their crosshairs, when previously they had shunned any consideration of mind. The impression is thus conveyed that neoclassical economics has sworn off its previous boisterous imperialistic tendencies and now is united in sweet concord with all the other social and natural sciences in developing a unified portrait of what it means to be rational. This brave new world of "behavioral economics" thus manages to square the circle of subsuming supposedly "irrational" agents under the continued sovereignty of old-fashioned rational choice theory, seemingly without breaking a sweat.

This pattern of historical narrative—one that takes complex questions of "rationality" as effectively reducible to the "rational choice" paradigm and its exploits—even extends to writers who might otherwise consider themselves critics of the rational-choice tradition. For purposes of illustration, we might resort here to Nicola Giocoli's book *Modeling Rational Agents* (2003). Therein he starts by acknowledging that in the first half of the twentieth century, neoclassical economic theorists had approached their model as a "system of forces" (and *not* ratiocination), but because they harbored a residual predilection to refer to minds, they were bedeviled by two strange (and unexplained) obsessions: to escape any reliance upon academic psychology of any stripe in all circumstances, and to eschew their seeming reliance on a presumption of "perfect foresight" of the future in their little agents. Those motives were internally inconsistent, insists Giocoli, and furthermore, could never have been reconciled on their own terms. Possibly relishing a sense of irony, Giocoli therefore argues that the staunch prophets of rational consistency were inconsistent in their own reasoning.

Luckily for economists, Giocoli intimates, some outsiders rode to the rescue. The deliverance from the dilemma was providentially provided by a few mathematicians and logical positivist philosophers, or so he claims, by reducing the meaning of "rationality" to the consistency of an empty set of formal relations—viz., what we have identified above as the formal axiomatic component of rational choice, restoring consistency to the inferentially challenged. The axiomatic method thus absolved them from any commitments to disciplinary psychology, while the marriage of probability theory and utility (attributed by all and sundry to von Neumann and Morgenstern) putatively absolved the agents from perfect foresight. Problems of knowledge were downgraded to problems of "risk." The new prophets of rationality proceeded to pronounce their project "rational" by fiat, and beyond all reasonable expectations, were stunningly successful in convincing others. The rest of the book then becomes an extended meditation on the history of game theory as one necessary consequence of this reconceptualization of rationality.

While Giocoli does strive to bring his argument to bear on many different texts, in the final analysis, in our view, his story just does not hang together. Consider the contemporary evaluation of Kenneth Boulding: "The epistemological theory of decision making is, of course, pretty empty unless we can specify ways in which the inputs of the past determine the present images of the future. Unfortunately, the observations of economists on this question [circa 1965] are for the most part simple-minded to the point of embarrassment" (1966, p. 7). The biggest complaint we might make is that the Giocoli narrative takes place in an amazing vacuum; other than the *deus ex machina* of the logical positivists, no attention is accorded to anything whatsoever happening outside the narrowly contrived coterie of neoclassical economists. Giocoli stumbles initially because he effectively sets out to write a history of decision theory (and not of price theory) over the course of the twentieth century, when in fact there was no such intellectual formation before WWII. As Kenneth Boulding said in his Ely Lecture to the AEA in 1965, "the Epistemological Question has received rather scant attention at the hands of economists."[23] Betraying the bad habit of reading current obsessions back into innocent predecessors, and presuming everyone in economics was naturally concerned with an ahistorical entity called "rationality," Giocoli imagines that his early prewar protagonists agonized persistently over the character of "the decision" in much the same way as they did after 1945—but there is no evidence of that. Indeed, if they shared much of anything in those halcyon pre-WWII days, rather it was a peremptory dismissal of "mind"; if and when cognitive matters did come up, they were dealt with as issues of "intelligence" (with the usual eugenic implications) rather than "rationality."[24] Simply having to resort to the similar mathematics of optimization and utility functions is nowhere tantamount to being driven by the same conceptual questions. The only way to begin to gain some deeper perspective on the elusive doctrine of "rationality" is to venture outside the narrowly conceived ambit of "economics" and ask: Where did the earlier neoclassicals derive their most nagging questions from, and what was happening in the other sciences that caused them grief and aggravation? That constitutes the core of the narrative in this current book.

Although much of this book covers the period after WWII, here we might venture a few brief observations about the prewar period, and its relationship to the attempt to render the history of economic orthodoxy as the history of rationality, through and through.

The story must begin with the fact that the original neoclassical model was copied from energy physics (Mirowski 1989). The early neoclassicals were by and large agnostic about mind: all that mattered for them was to equate the formalisms of energy with something they diffidently called "utility," so that they could portray the market as deterministic and as law governed as the rolling of a ball to the bottom of a bowl. First and foremost, the neoclassicals were drawn like moths to the flame of science; very few of them were acolytes of "utility" as a serious theory of mind. Only a very few card-carrying mathematical neoclassical economists of the first three generations could be bothered to follow up on the mathematical metaphor and ask what it may have explicitly implied for psychological predispositions; and those inquiries were ignored by the newly professionalizing cadre. It is revealing that Giocoli opts to call this a "System of Forces" approach to economics—somehow he stops short of the full nine yards and the admission that it was merely a bowdlerized physics. Furthermore, the physics inspiration reveals why "perfect foresight" was not the dread albatross that Giocoli conjures for the prewar era: the theory was purely static, as nearly every neoclassical economist freely admitted; and moreover, many realized that classical mechanics itself had no need of perfect foresight, since it was purely time reversible.

One of the quaint characteristics of prewar neoclassical economics was that it was concertedly backward oriented: events in the past were thought to determine realized prices and quantities in the present: think of the Austrians and their capital theory, or the endless Marshallian fascination with cost structures inherited from the past. Causality was assumed to obey time's arrow. It was only after WWII that neoclassical price theory swung 180 degrees on the time axis, with occurrences in the future supposedly governing current choice and decisions. This rotation on the time axis is one of the main epoch-making events we explain in this volume. "Information" was concerned primarily with what had not yet happened—not, demonstrably, about imperfect knowledge of the past. What Giocoli paints as a source of deep anguish we argue was merely a minor distraction for neoclassicals before the World Wars.

However, we concede it is demonstrably the case that neoclassicals have always sought to absolve themselves from the external strictures of formal or professional psychology. If there has been a constant in the history of neoclassical economics, then that is surely it. It also turned out to be the necessary prerequisite for the existence of an economics of information. It is curious that Giocoli could not glance at what had been happening in psychology at the turn of the twentieth century, for then he might begin to grasp how very distressing the "science of the mind" was trending for the early neoclassicals. The new, fascinating thing on the continent back then was Freud and psychoanalysis; the enthusiasm in America was "habit psychology." In both cases, the preferred stance in most social sciences was to stress the great extent to which the "irrational" governed people's behavior: conscious rationality was portrayed as a weak, leaky vessel tossed about on a roiling set of instincts, urges, and inaccessible unconscious. Indeed, most of the early neoclassicals did not characterize their theories as expressions of generic rationality, for the very important reason that the median attitude in the vanguard social sciences circa 1920 was that the great mass of humanity was not very rational, the proud breakthrough of the social sciences having had the courage to embrace the pervasive yet bitter truth of this stricture.[25] Indeed, in the alternative narrative we shall shortly explore, this stood as the premier "fact" of prewar social science. Thus, to promote their nascent project as the construction of a theory of "rational choice" would have been extremely quixotic, if not suicidal, at least prior to the 1940s; discretion being the better part of valor, most neoclassical theorists did not go there. The least discreet among them, Vilfredo Pareto in his *Manuele*, did suggest that economics was the province of "logical" action, whereas sociology was stuck with the "non-logical" dregs; but even he immediately got tied up in knots over where this dividing line could be drawn: "non-logical action does not mean illogical; a non-logical action may be one which a person could see, after observing the fact and the logic as the best way to adapt the means to the end; but that adaptation has been obtained by a procedure other than that of logical reasoning."[26] In claiming the mantle of "logic" for neoclassical economics, Pareto in no way was asserting that the great mass of people adhered to logical behavior—far from it.

By and large, neoclassical economics propounded neither a full-blown theory of "choice" nor one of "decision" in the prewar era. That is the first clue that the raft of "internalist" histories and vindicationist accounts are deeply misguided when it comes to understanding the rise of "rational choice theory." And if that be the case, then what was it that caused the watershed around the mid-twentieth century, the rise of the "decision" as the hallowed hallmark of our (in)humanity, a tremor of tectonic proportions that Giocoli detects as well. Was it the handiwork of the nefarious "positivists"? Not by a long shot. The "billiard ball" model of rational choice came from outside economics—but where?

The short punchy answer, fleshed out in this volume, is threefold: it was the military, the rise of the digital computer and its complement "information," and last but not least, the rise of the political doctrine of neoliberalism.[27] These assertions may perhaps strike the reader as curious, except for the fact that they constitute the thematic core of much of the new historiography, represented by Erickson et al.'s *How Reason Almost Lost Its Mind* (2013) and Hunter Heyck's "Producing Reason" (2012). Starting with Heyck:

> The core of my argument is that social scientific discourse about choice from the 1920s to the mid-1970s was part of a discourse about reason and the prospects of democracy.... [It] was a novel blend of pessimism about the scope and quality of human reason and optimism about the power of social

and technical mechanisms for producing rational choices.... Instead of asking whether people were rational creatures, the question should be, what is the best system for producing rational choices? *The object of study needed to be the choice, not the chooser.*[28]

This reification of the choice independent of the consumer in economics could only happen with a commensurate development of equal import: the reification of information independent of the cognizer. This dual disembodiment could not have occurred without a further catalyst: the rise of the electronic computer. Here was a machine that, at least in the estimation of some of its most avid promoters, could *think*; and it accomplished this by serving as a factory for the production of information by means of information. Nothing better exemplified the reification of "choice" than the manifest "rationality" of a nonhuman chooser. The American military notoriously served as the incubator for the modern computer with its von Neumann architecture; what has only more recently come to be appreciated was that the American military was also the incubator for modern "decision theory." As Heyck (2012, p. 110) reports, "If one searches an online database such as JSTOR for articles from the 1950s having to do with decision-making, one is over 90% likely to find that the author of the piece was at least partially sponsored by the Office of Naval Research or RAND." The authors of *How Reason Almost Lost Its Mind* second the significance of this connection: "Key to the contrast between Enlightenment and Cold War rationality is the rise of the modern automated algorithm in conjunction with the economic rationalization of calculation.... [T]he rule-following computer appears throughout this history as point of reference for exploring rational conduct."[29]

Note well that individual agency has been effectively downgraded in these histories, in tandem with the new genre of spy fiction. If we have any reservations concerning this newer vintage of history, they have to do with how the evidence has been presented and framed. First, one thing that strikes the reader of these narratives is that they never actually explore the models of the economists in any detail. To bring the point home that we now live in a radically different epistemic era from that of our forebears, there is nothing like observing the culture of disembodiment inscribed in the mathematics of the economic models themselves. We strive to rectify that oversight in the chapters that follow. Second, these histories often mention that cold war anxieties involved worries over the efficacy of democracy, but never delve into the proximate sources of the hostility to democracy embedded in the political economy that drove the microeconomics and decision theories. Here is where a richer understanding of the rise of neoliberalism can reveal how the economists were partly at the mercy of larger political forces, ideological cross-currents that they could only incompletely comprehend. We return to this phenomenon in our final chapter. Third, to reprise our original complaint, in the final analysis, what we provide here is not really a history of human rationality so much as it is a history of the rise of the information concept as the pivot point around which economics, computation, and politics rotated in the twentieth century, only to wander off in an unanticipated precession. The ultimate inversion of conventional narratives is to realize that knee-jerk humanist concerns about what human beings are *really* like and how humans *really* think have become all but inconsequential in modern economics. The residual inclination to drape existing histories around the vanishing cold war notion of "rationality" can therefore only obscure our modern situation.

The tendency to write the history of modern economics as a history of "rationality," hence, misses much of the real action in the intellectual history of economics among the sciences. Some methodologists can trumpet that we live in a new pluralist age of orthodox economics only by ignoring the anti-humanist undercurrents of economic ideas.[30] Others may praise the new entrepreneurial spirit of the economics profession without the slightest notion of what it implies. We are determined to write the history in a different register. It is far better to track the peregrinations of the entity called "information" as it has wrought havoc in the evolution of modern microeconomics. But to embark upon that journey, it is first necessary to have a brief introduction to the history of the concept of "information" in the larger culture.

From a brief primer on information, we turn to explore where economists first managed their rendezvous with it, and we discover, to our surprise, that it happened initially with the Neoliberal Thought Collective and Friedrich Hayek. The response to this incursion was mounted by the orthodoxy at the Cowles Commission, leading to at least three distinct model strategies. But the logic of the models led to multiple cognitively challenged agents, which then logically led to a stress on "markets" to rectify those weaknesses. Unwittingly, the multiple conceptions of agency led to multiple types of markets; and the response of the orthodoxy was to shift research away from Walrasian themes to what has become known as "market design." But internal contradictions in the design program led to a startling conclusion: just like their agents, the orthodox economists turned out to be not as smart as they had thought. A little information had turned out to be a dangerous thing.

[2]

THE STANDARD NARRATIVE AND THE BIGGER PICTURE

> Economies are built out of information. This has been true from the Stone Age to our knowledge economy today.
>
> Eric Beinhocker[1]

As explained in the previous chapter, we start from the premise that economists possess no special expertise in the mysteries of human nature, or at least no more than the man on the street. Rather, the prime directive of professional economists has been the attainment of *scienctific* status for the field; and what this has meant after World War II is that they must participate in the most important trend in the modern sciences—namely, that they have elevated *information* to the pinnacle of their theoretical structures. Hence much of what has been misunderstood or misconceived as controversies about "rationality" have been, in fact, struggles over how to insinuate "information" into the canonical model of microeconomics, which dated back to the 1870s.

The fact that "information" has transformed the natural sciences from the 1940s onward is a proposition now widely taken for granted among scientists and historians. Some of the more philosophically inclined, dissatisfied with a basic trinity of matter, energy, and information, go so far as to attempt a further reduction of all three into the One True Entity.² We need not defer to its enthusiasts to that extent in the current context; it is enough to paraphrase the popular historian James Gleick, who said that after the 1940s, information was the axis around which the entire world began to spin.³ While we will only touch upon a few small facets of that massive transformation of the natural sciences, the timing of its appearance is of paramount importance. That information revolution preceded the watershed event that many media theorists take for granted—namely, the appearance of the Internet and everything now associated with it. The key period turns out to have been the years surrounding World War II, and it is from there our own narrative will soon begin.

Over and above the issue of timing, it is indispensable to realize that recourse to "information" in neoclassical economics could never have been a matter of effortless appropriation. Inevitably, it could never have been simple larceny, if only because there was no simple, unique thing to appropriate. Information had proved that it "can add colors to the chameleon, / Change shapes with Proteus for advantages, / And set the murderous Machiavel to school."⁴ This multiplicity was only one of the many ways the history starts to look different, once we situate the advent of "information" at its center. One rather obvious effect is that we begin to glimpse that "rationality" remains a rather empty concept, at least until one begins to have some idea what a theorist means when she suggests an agent "knows" something. Another eye-opener comes in noticing that the referent of information in economics changes appreciably over time, largely in reaction to changes in the cultural themes swirling around theories of information and knowledge. Perhaps the most bracing shock consequent upon shifting the frame is a dredging to awareness of the gradual transition from economists' regarding information as an unalloyed good to praising ignorance as the appropriate state of a dedicated market participant. In what brave

THE STANDARD NARRATIVE AND THE BIGGER PICTURE

new world can the economist Gary Gorton unself-consciously write: "We show that preserving symmetric ignorance in liquidity provision is welfare maximizing and strictly dominate symmetric and even perfect information."[5] Or, to paraphrase, blessed are the meek and stupid, for they shall promote liquidity and stability.

Never before have the high priests of efficiency been so happy to proclaim that a sucker is born every minute. Economists nowadays hold it as gospel, but must not let on that is so. One major motivation for writing our book is to provide a meditation upon why and how such an incongruous trajectory has pirouetted to meet its antithesis. Why must an economics of information end up propounding a doctrine that economic agents should be ignorant? While the causes are many and the detours abundant, there is one profound philosophical question that needs to be broached before we set out on our narrative.

There is a bit of commonplace wisdom found in the community of generalist intellectual historians that the era of the 1970s–1990s was one of contextualization and historicization of ideas, as well as the exploration of various formats of "identity politics." David Hollinger, an intellectual historian and a major proponent of this view, summarizes the trends in a half-facetious way as "Kuhn, anti-racism, feminism and Foucault." Whatever one might think about such broad-brush generalizations, he also explicitly insists that the economics profession was uniquely immune from the entire trend:

> One could write an exceedingly detailed history of this entire episode without mentioning a single economist. This exclusion applies even to those economists who sometimes write about large and hard-to-solve problems for transdisciplinary audiences, such as Robert Heilbroner, Albert Hirschman or Amartya Sen. . . . *Epistemic universalism was never seriously challenged.*[6]

One might be inclined to agree that most economists never actually gave the feminists or Foucault the time of day (although it appears many were assigned Kuhn some time in their student years, only to cement their belief that economics had attained "normal science" status), but it is a serious misunderstanding of the historical record to accord the economics discipline immunity from cultural trends in epistemology. The reason Hollinger and his colleagues have missed this connection is that economic arguments about identity and truth were played out in a space somewhat removed from the generalist magazines and newspapers that often provide such historians with their major sources of evidence. We shall demonstrate in this volume that economists were also swept up with popular concerns that truth may appear to be contingent and dependent upon personal identity during the same period, but that those disputes were carried out primarily in technical discussions of mathematical models in specialized journals; furthermore, the issues of identity that drew the most attention were not those of race and gender but, rather, the minatory perseverance of an epistemic divergence between the professional economist and the quotidian agent, with lesser (but not unsubstantial) attention paid to epistemic divergences between agents.[7]

The disparagement of the sagacity of *Homo economicus* has direct bearing on the asymmetry between agent and analyst. There might be a sense in which the legitimacy of the modern economics profession almost demands that their representative rational agent needs to be rendered less knowledgeable (or, to court offense, more stupid) than the orthodox economist himself. To allow the agent to know substantially more than the economist is transparently a nonstarter, since in that instance the pretense to special economic expertise comes to naught. One might then suspect that the sweet spot is for the agent to know exactly everything the analytical economist knows—one observes this in modern assertions that the agent "knows the true model of the world"—which conveniently turns out to be identical to the orthodox neoclassical model—but that option also has its own severe drawbacks. These include the overwhelming evidence of empirical violation of rational choice models; but more to the point, there are the compromised ambitions of the economist to provide privileged normative prescriptions for the economy. If everyone else happens to be situated on the very same epistemic footing, then the man in the street really has no reason whatsoever to attend to economists and their prognostications; and, quite starkly, economists have nothing relevant to impart to them.[8] Can the student appreciate that, for neoclassical economists to hold their heads high in society, and believe they can model information and its consequences, it is imperative their imagined representative agent be stupider than the median economist?

The repercussions might even be momentous. What happens when the person on the street comes face to face with this kind of scorn (if that ever comes to pass)? Will the contempt of the economist be met with a different sort of disdain by the public?

WHAT EVERYONE KNOWS

Once upon a time—say, around the era of David Ricardo and Karl Marx—political economy was primarily concerned with the production of national physical wealth. This "classical" notion tended to hang on long into the twentieth century, well after the invention of neoclassical economics in the 1870s. Nevertheless, there was no denying that within neoclassical economics, notions of exchange had displaced those of tangible production as the primary topic of interest; this informed the definition of economics articulated by Lionel Robbins that its proper subject was the "allocation of scarce means among given ends." But subsequently, something rather extraordinary happened around the middle of the twentieth century, gaining momentum as the century waned. More and more, economics at the cutting edge (as opposed to the textbooks) became relatively cavalier about treating trade as static allocation, and instead became all wrapped up in the image of The Market (or else the agent) as a processor of information or knowledge.

Appeals to information and knowledge pop up almost everywhere these days: the efficient markets hypothesis in finance, common knowledge in game theory, rational expectations in macroeconomics, asymmetric information in principal/agent theory, adverse selection in mechanism design, focal points in behavioral economics, and so on. For many economists, these elaborate constructions of the market consequences of information are the main reason the modern economic orthodoxy is always an improvement on your father's economics.[9] We provide the following paean to modern success, lifted from a recent blog:

> It is a strange fact that many social scientists feel economics to some extent stopped progressing by the 1970s. All the important basic results were, in some sense, known. How untrue this is! Imagine labor without search models, trade without monopolistic competitive equilibria, IO or monetary policy without mechanism design, finance without formal models of price discovery and equilibrium noise trading: all would be impossible given the tools we had in 1970. The explanations that preceded modern game theoretic and information-laden explanations are quite extraordinary: Marshall observed that managers have interests different from owners, yet nonetheless are "well-behaved" in running firms in a way acceptable to the owner. His explanation was to credit British upbringing and

morals! As Stiglitz notes, this is not an explanation we would accept today.[10]

People with advanced training in economics seem to think they know what this transformation has been about and why it has happened. One indicator might be that "information" was added as a subject category in the American Economic Association subject taxonomy in 1976; this in itself suggests a postwar heritage. As Kenneth Arrow once put it: "one of the biggest differences between 1950 and 2000 is the much greater role now given to . . . knowledge and information. . . . Nobody would have denied the importance of these in 1950, but the tools to handle them did not formally exist then."[11] But, in fact, one could just as easily, and with justice, assert the opposite: certainly we possess many more "tools," but the impression that clarification followed formalization needs to be approached with a modicum of salt. Interestingly, in the interview, Arrow did not bother to identify even one specific hallmark doctrine within the orthodoxy that owed its existence to the stipulated sea change. Further, Arrow hedged his bets by citing "knowledge *and* information," and this begins to reveal the profound ambivalence and unapologetic befuddlement that often besets anyone attempting to make sense of the history of this literature. Rather than concede the distinction between knowledge and information as meaningful or significant for economics, Arrow rather haughtily responded to a request to clarify the distinction with the dismissal, "I am afraid the topic does not inspire me."[12] The Muse of Information moves in mysterious ways for neoclassical economists.

Those fortunate enough to have experienced Philosophy 101 will come away wary of people who conflate knowledge and information with abandon, much less defer to those who treat knowledge as if it were something that needs no definition, since everyone who thinks he or she has it, does indeed have it. Information enthusiasts outside of economics manage to signal their awareness of the importance of some crude distinctions: "Information can be moved around easily in the products that contain it . . . but knowledge and know-how are trapped in the bodies of people and networks that these people form."[13] It seems fairly commonplace to concede that knowledge is something not readily reified, and yet many economists still seem to resist this insight. Indeed, a historical wariness toward this tendency is one of the sentiments we hoped to evoke indirectly by means of this book's title. In this regard, the oft-neglected Kenneth Boulding was a far better guide to these issues than Kenneth Arrow:

> Knowledge, however, has a dimension which goes beyond that of mere information or improbability. This is a dimension of significance which is very hard to reduce to quantitative form. Two knowledge structures might be equally improbable but one might be much more significant than the other.[14]

Boulding, somehow, did not make it into the modern economist's Hall of Fame.

The prudent intellectual historian cannot help but be struck by the fact that when some judicious souls began to write surveys of this newfound enthusiasm among economists for information in the 1980s and 1990s onward, they came away bearing impressions of something akin to the Tower of Babel.[15] One such attempt at a history published in 1995 began by quoting George Akerlof on the initial rejection of his famous "Lemons" paper: "They were afraid if 'information' was brought into economics, it would lose all rigor, since in that case almost anything could be said—there being so many ways that information can affect an equilibrium." The authors then concluded their survey with: "we encounter an over-abundance of results and/or equilibria; almost anything can happen."[16] "Anything goes," on its face, would hardly seem the prescription for a successful intellectual tradition.

One could confront the impressions of disarray in the historical archive in many different ways, but our task here is to set out to tame the blooming buzzing confusion by getting a clearer idea of what the economists did and did not accomplish. First off, we need to establish that there really was something new under the sun in economics back in the twentieth century. As Dan Schiller perceptively queried *over a quarter century ago*: "Why wasn't the status of information a major topic in economic theory in 1700, 1800, or 1900? Why was it only in the postwar period that the economic role and value of information took on such palpable importance?"[17] The easy retort—that it was absent-mindedly "overlooked," exiled to peripheral vision until WWII—simply will not wash. Neither will the bad habit of searching for precursor statements from previous eras, anticipations which were never really there. Rather, we shall insist that there were specific historical conditions in the 1940s that stood as necessary prerequisites for the seeming inevitability of the juggernaut of the economics of knowledge. But the era from the 1940s to 1970s is a blind spot for the modern economist—at least when it comes to the economics of information.

What manner of impressions about the history of the economics of information does the modern student receive from her theory teachers, even if only as capricious stage-setting preliminaries? More often than not, there she encounters some assertion that nothing really happened before the 1960s or 1970s. Some may cite George Stigler's (1961) paper as kicking off the literature, or else that of William Vickrey (1961), and perhaps the aforementioned 1970 "Lemons" paper by Akerlof (1970). What is noticeable about these supposed landmarks is that, by and large, their actual models turn out to be unrelated to pretty much anything that follows in the curriculum our student undertakes. For instance, Stigler's sequential search for lower market price among dispersed differentiated purveyors has been pretty much forgotten. Instead, the class in the economics of information mostly assumes one of two formats: either a meandering survey of picky small variations on the benchmark von Neumann–Morgenstern individual expected utility model to capture something called "uncertainty"; or if the class is skewed toward Nash game theory, then the topics tend to revolve around how agents seek to mislead, dominate, and otherwise confuse their opponents, with topics called "asymmetric information," "cheap talk," "adverse selection," "private values," "signaling," "moral hazard," "common knowledge," and the like. The macroeconomists are partial to the former tweaks to decision theory, whereas microeconomists—and particularly market designers—favor the latter pastime of Liar's Poker. Over time, it seems the game theory advocates may have come to numerically dominate the decision theorists, and as that happened, their own favored potted history has grown more austere and sparse, ignoring the vast mass of information theorizing of that era in favor of their one purportedly canonical general model of information, the Bayes-Nash model first propounded by John Harsanyi in 1967. In both cases, the contested notion of "rationality" keeps getting amended willy-nilly to supposedly handle information, while past approaches get erased. There is a third genealogy of the economics of information that tends to be favored in business school classes outside of economics departments proper, especially in finance, which is instead structured around the efficient markets hypothesis: in that potted history, it is The Market itself that stands as a superb processor of information in its own right. There, cognition and rationality are dispensed with altogether.

First, even at this most superficial level, we can observe that there has been no single canonical model of the economics of information at almost every step along the way. The student haphazardly gets arbitrarily schooled in one manifestation or another, depending on the angle of approach to their own specialization. Hence, the conventional notions of the history of information in economics tend to be a little incoherent. Furthermore, the potted orthodox histories turn out to be singularly uninformative about the deeper issues driving this need to privilege "information," or perhaps "knowledge," in economic theory over time. For instance, why would someone think the unique best way to approach the question of knowledge about the economy was through an inductive definition of "uncertainty"? The words sound like they have a family resemblance, but it doesn't take a philosophical genius to realize they are hardly identical. As Thomas Schelling once observed back in 1962, "There is a tendency in our [theories of] planning to confuse the unfamiliar with the improbable" (p. vii). One wag later said that you can always tell a decision theorist because of his struggles with the undecidable: he seemingly can't tell the difference between irrationality and ignorance.

Second, existing orthodox histories of economics do a poor job of explaining what *effects* the economics of information have wrought. Because we do not automatically share the orthodox obsession with "progress," it will be crucial to establish as historians whether or not this *nouveau* fascination with knowledge actually has thoroughly reconstructed the heartland of orthodoxy, or instead, as so often happens, merely clutters up the periphery with sprawl. Here again, there has been no general consensus on this issue among the orthodox. At one end of the continuum, Joseph Stiglitz has hinted from time to time that the previous world of neoclassical economics was turned upside down by the economics of information:

> There is no single new Law of Economics.... The world is not convex; the behavior of the economy cannot be described as

if it were solving a (simple) maximization problem; the law of supply and demand has been repealed.[18]

We know of hardly any orthodox economists today who would subscribe to any one of those precepts. Conversely, at the other reach of the continuum, various respected economists have hastened to reassure the rest of us that nothing has really changed after all, and that, contrary to most impressions, the ongoing info-fascination was just a minor variation on the age-old wisdom of neoclassical economics.[19]

Both narratives turn out to be equally implausible; and this ushers us toward the nub of the present philosophical conundrum. How can it be that so many economists have come to believe there has been some sort of Great Transformation of Economics into a Science of Knowledge in the last half-century—the dawn of a higher wisdom buttressing an information economy—and yet be utterly incapable of producing even a sparse, clean consensus on the hallmark doctrines of the New Order?[20] Why are they so convinced that "information" constitutes the panacea for so many of their problems? Could it be due to the internal logic of their economic model, or has the pressure come primarily from outside the discipline?

Risking ridicule, or maybe just a standard paradox of self-reference, in this book we seek to pose the question: How do these economists *know* that knowledge has become ineluctably central to their discipline? This paradox will lead us to an even thornier question: How is it possible that a neoclassical economic theory, committed to a thoroughly ahistorical, noncontextual theory of equilibrium (and notoriously weak on how that equilibrium is attained), could provide an adequate account of the process by which knowledge is gained, interpreted, and understood? Can the dynamic character of knowledge and interpretation be anything more than a square peg forced into the round hole of given preferences?

Of course, we come equipped with no magic Philosopher's Stone to settle these thorny questions once and for all; nor would we ever want to. But we do think we can do a better job in organizing the history of economic thought around information, as a prelude to a coherent answer. We will therefore avoid the standard practice of structuring narratives around all those various "applied" areas of economics (e.g., macroeconomics, finance, decision theory, signaling, and adverse selection); rather, we divide the story into the different ways that information and knowledge themselves have been conceived and elaborated though time. In other words, rather than dutifully following the economists in their arbitrary sequences of applied models, we will adopt a more *epistemic and ontological* approach. The questions that will serve as our protocols will be:

- Who or what is supposedly doing the thinking?
- Is there any real cognition going on?
- What, precisely, qualifies as information and where did the model inspiration come from?
- Have informational considerations been effortlessly grafted onto the prior neoclassical model of fixed utility functions and fixed endowments, or has the bequeathed canonical model become compromised over time?

These questions will lead us to argue that the older, potted, orthodox histories of the economics of information have been misleading at best, and at worst have been utterly confused in some of their more commonplace variants. Our alternate narrative should help the student understand how the profession has arrived at this curious impasse, where ignorance of their own history has permitted economists to believe the most extraordinary things about knowledge and their place in it, without any serious consensus among their peers.

And more to the point, fortified with this history, the student will be much better prepared to recognize the possibility of a New Economics, if and when she encounters it. She may even experience the joy of learning something.

[3]
NATURAL SCIENCE INSPIRATIONS

A century ago, "information" did not have much cultural resonance as a concept. It was a nondescript word: "An item of training; an instruction." Yet we now find ourselves poised and pinioned in the Information Age. Which, by the way, the *OED* defines for us in its dry prose: "the era in which the retrieval, management, and transmission of information, esp. by using computer technology, is a principal (commercial) activity."[1] Its original referent had been derived from the ancient Latin precursor: the verb *informare*—to give form to; to shape; to mold. Information at its birth was the act of *infusion with form*. Where, and how? In the beginning, the forming takes place *in the mind*. But a big part of our narrative is the way in which the verb got reified into a noun, first into a number, then into a thing, and finally, into a cosmic principle of organization around which our age putatively revolves.

As we would expect from the broad outlines of the history of economics, the truly earth-shaking innovations did not originate from within the discipline but, rather, tended to trace their inception back to the natural sciences. Through a sequence of steps far too labyrinthine to trace here,[2] late nineteenth-century developments in thermodynamics first stabilized the key concept of entropy; then a twentieth-century concern with noise in circuits and communication channels (especially at Bell Labs) led engineers to equate entropy with a measure of something they called "information." This version of "information" became a topic of explicit mathematical models starting in the 1920s, and had begun to spread throughout the *natural* sciences by World War II. Claude Shannon's 1948 "information theory" started off as an attempt to theorize cryptography, but was later published in the open literature as a theory of the capacity of channels to convey messages.[3] In the technical context, Shannon proposed a measure of the difficulty of sending a set of symbols from emission point A through a communications channel beset with noise, to a receiver situated at point B, as shown in figure 3.1.

Shannon's innovation was to treat the string of symbols conveyed as a stochastic process, with each symbol possessing its own characteristic probability. In one fell swoop, this move rendered the semantic aspects of communication utterly irrelevant for the theory—a point which will soon assume some significance. Shannon then posited that the mathematical expression for the average *improbability* of a string of such symbols would be exactly the same as the earlier definition of physical entropy:

$$H = -\sum p_i \log_2 p_i$$

However, in this instance, the probabilities referred to the reconstitution of symbols in a receiver, rather than states of physical position and velocities. Shannon had recast communication as the

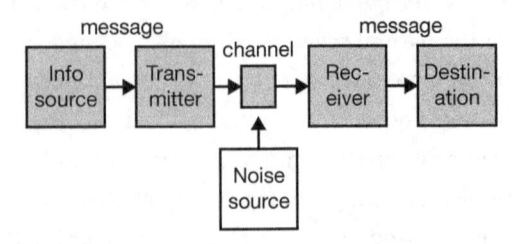

Figure 3.1. Shannon Information Theory.

selection of symbols from some preset fixed roster, rather than the stipulation of anything truly novel or unprecedented. In the simplest possible case, with a binary alphabet and equiprobable symbols, $\log(½) = -1$, and $H = 1$; this unit was named the "bit."

Something marveled at by almost every historian of science who has examined this train of events was the alacrity with which almost everyone beyond the communication engineers immediately forgot or ignored all the relevant qualifications concerning the formal Shannon information concept—its emptiness of meaning, its predicate of a fixed transmission channel, its curious construction of a stilted notion of "uncertainty," and its crude mechanical inspiration—and proceeded to herald a new dawn of a science of information within their own fields.[4] One observes this as early as the Macy Conferences on Cybernetics; in 1951, Shannon was already cautioning the conferees that his information concept was not what they were imagining it was.[5] Undeterred, "information" soon became all the rage in postwar psychology, particularly the decision theory variant, although the enthusiasm burned out there by the 1970s. Biologists were often serial offenders, mixing and matching information metaphors with their newly discovered DNA.[6] The famed mathematician Benoit Mandelbrot (1953) began his career by linking Shannon information to game theory. Fritz Machlup bemoaned that "the failure to find, and perhaps the impossibility of finding, any ways of measuring information in this ordinary [vernacular] sense induced many to accept signal transmission, channel capacity or selection rate as a substitute or proxy for information."[7] In other words, intentionally or not, Shannon information had been taken as warrant to posit it (or something similar) as a measure for whatever that particular school of research wanted it to be, and nowhere was the practice embraced with more gusto than in economics.

Scholars in psychology were getting skittish about this by the mid-1950s:

> There is something frustrating and elusive about information theory. At first glance, it seems to be the answer to one's problems, whatever these problems may be. At second glance it turns out it doesn't work out as smoothly or as easily as anticipated.... So nowadays one is not safe in using information theory without loudly proclaiming that he knows what he is doing and he is quite aware that this method is not going to alleviate all worries.[8]

Machlup, coming from his Austrian neoliberal background, was appalled. "Information has become an all-purpose weasel word," he sputtered in 1983; people would indiscriminately shift between denoting a process of communication, a symbol string, a semantic content, an index of "uncertainty," an object independent of any mind, a stock of previous transmissions, encoded instructions, and worst of all, a synonym for "knowledge." Vague similarities with negentropy prompted a few social scientists to dabble in "negative information," a very frightening concept, but one not to be confused with ignorance. Knowledge had conventionally referred to a cognitive state of being; it might be personal or it might be social, but certainly was not something disembodied from human life, at least prior to 1980. "Data data everywhere, and not a thought to think," Machlup growled. Yet the trend over time, especially after the 1980s, was to leach "knowledge" of many of its original connotations, and conflate it more and more with "information."[9] The computer was not the only culprit in this respect. After all, biological reductionists were suggesting human identity was potentially reducible to DNA; science fiction and artificial intelligence mavens regularly speculated about downloading "yourself" onto some computer disk and thus freeing yourself from corporeal integuments. The human soul was growing more wispy and elusive, whereas information just seemed to grow more robust and pervasive.

Of course, Shannon information theory was not the only font of legitimacy for the turn to scientific formulations of cognitive matters and communication glitches after WWII. A concern over military intelligence led to the founding of a science of military decision theory, carried out under the rubric of "operations research" and the newly minted field of "communication studies."[10] Military mandates often sought to detach "information" from content and semantics. Many of the many curious hybrids funded by the military may seem quaint in the era of the Internet: Project Revere, a University of Washington enterprise in the 1950s, which sought to study "the effects of the flow of information" through communication networks that diffused messages beyond targeted recipients, actualized this by conducting leaflet drops from airplanes on thirty-five American communities that had previously been constrained by military order neither to forewarn the public nor provide official comment upon these unprecedented events.

These exercises in the sciences of war and their applications were crucial for subsequent developments in all the social sciences, and were full of implications for conflation of the ontological status of "information" versus "knowledge," but for purposes of brevity we must merely take them as given here. Likewise, there were further developments growing out of military patronage that we should highlight, but must of necessity neglect provision of their own historical narratives, including: (a) the elaborations of mathematical statistics and game theory; and (b) the construction and stabilization of the first electronic digital computers in WWII.[11] Both projects started outside of economics, but proved irresistible as they swept the postwar sciences. The physical instantiation of "machines who think" (whatever was intended by such locutions) provided irresistible metaphors for cognitive activities, many of which would be taken up to various degrees in all the social sciences in the postwar expansion of academic research. And, there would be no field of "artificial intelligence" without the computer. These physical traditions concerning the sciences of information and its processing set the stage for a later reconstruction of economics, rendering some attempt at accommodation nearly inevitable.

[4]

THE NOBELS AND THE NEOLIBERALS

There is a genre of history that is written unreservedly from the perspective of the winners. And manifestly, there are no more prominent winners in the orthodoxy than the recipients of the misnamed Nobel Prize in Economics. (The correct designation is the Bank of Sweden Prize in Economics in Honor of Alfred Nobel.[1]) One theme of this chapter is that the sequence of the Nobels does not provide a reliable chronology for the rise of the economics of information.

We provide own select roster of winners in table 4.1, consisting of the subset of laureates in the first twenty-five years of the Bank of Sweden Awards who had significant input into what subsequently became orthodoxy in the economics of information/knowledge. While beginning to mark the initial boundaries of who or what eventually fell into that contested category, the table also reveals something previously quite unnoticed in the history of thought: five of these nine laureates in that period were members of the Mont Pèlerin Society (MPS), a society founded by one of those laureates, Friedrich Hayek.[2]

The fact that roughly half of the recipients would just happen to be members of any one single club, much less an organization that recruits and vets its members instead of allowing them to freely join, and whose roster has *never* exceeded 500 persons, would seem

Table 4.1 ECONOMICS OF INFORMATION/KNOWLEDGE NOBEL LIST, 1969–1994

Year	Laureate	Country	Rationale
1970	Paul Samuelson	United States	Properly anticipated prices are random: origins of efficient markets hypothesis
1972	Kenneth Arrow	United States	*Economics of Information*, vol. 4 of *Collected Papers*
1974	Friedrich Hayek*	United Kingdom/ Austria	Innovates notion of market as information processor
1978	Herbert A. Simon	United States	Bounded rationality; artificial intelligence
1982	George Stigler*	United States	Market prices as subject to sequential search
1987	Robert Solow	United States	Technological change as mathematical residual in growth theory
1988	Maurice Allais*	France	Attack on expected utility theory
1991	Ronald Coase*	United Kingdom	Discovery and clarification of the significance of transaction costs and property rights for the institutional structure and functioning of the economy; attack on state-sponsored communication outlets like BBC
1992	Gary Becker*	United States	Human capital theory

*Member of Mont Pèlerin Society

to be a curiosity that would have demanded some explanation. The short explanation that we shall access here is that the MPS served as the command center during early phases of the establishment of neoliberal doctrine; the crucible of neoliberalism, in turn was the prime incubator within modern political economy responsible for elevating the status of information to its predominance in the later twentieth century.[3] This was true whether the economist in question shared the neoliberals' politics or was firmly opposed to them.

Why does the student interested in economics and its relationship to information need to know something about the Mont Pèlerin Society? Because lacking that, it will prove impossible to fully comprehend the outlines of our alternative history of economic thought. For those who think of the economics of information as beginning with Kenneth Arrow or George Stigler or Joseph Stiglitz or Paul Romer or Nash game theory, we aim to convince them that they are misinformed.[4] The rise of a self-conscious "economics of information" really starts with Friedrich Hayek and the Mont Pèlerin Society; but that proposition would seem garbled and obscure in the absence of (at least) some acquaintance with a basic primer on neoliberalism.

The Mont Pèlerin Society was first convened in 1947, to rethink and recast older classical liberalism in a set of doctrines more suitable for the world after the Great Depression and WWII. It did not consist solely of economists, but also included philosophers, other social thinkers, and not a few rich businessmen. In their discussions, which were often closed to outsiders, they hashed out a set of political doctrines that would rescue the world from what they considered to be the onset of a socialist disaster. The project has never been considered finished, and the MPS continues to meet regularly, down to the present.

Neoliberalism as a body of thought itself did not exist before World War II, and is only now beginning to be the subject of scholarly research in the history of economics.[5] The most important fact about the neoliberals, at least after they got over their earlier interwar interval of disarray, was that they were *not* advocates for a simplistic laissez-faire: that is one of the crudest canards concerning the contours of late twentieth-century political economy. Some of its opponents have called it "market fundamentalism"; but that also misses the point. The goal of the neoliberal project has been to redefine the shape and functions of the state, not to hobble or destroy it. In the interests of brevity, we will list a subset of *six* principles of neoliberalism developed jointly and severally by members of the MPS during the last half of the twentieth century, relevant to our history. These six precepts nowhere exhaust the intellectual innovations of the MPS. Novices should be warned the six principles are merely a glimpse of a much more elaborate political theory, to which we cannot do justice to here.

SIX IMPORTANT TENETS OF NEOLIBERALISM

1. What sort of "market" do the neoliberals want to foster and protect? While one wing of MPS (the Chicago School of Economics) has made its career attempting to reconcile one version of neoclassical economic theory with neoliberal precepts, other subsets of the MPS have innovated entirely different characterizations of The Market. The "radical subjectivist" wing of the Austrian School of Economics attempted to ground The Market in a dynamic process of discovery by entrepreneurs of what consumers did not yet even know what they wanted, owing to the fact that the future is radically unknowable. Perhaps the dominant version within the MPS (and later, the dominant cultural doctrine) emanated from Friedrich Hayek himself, wherein *"the market" is posited to be an*

information processor more powerful than any human brain, but essentially patterned upon brain/computation metaphors.[6]

This is a key point in the history of the economics of information. This doctrine had three distinctly different interpretations in Hayek's own career alone (discussed in chapter 6), but is often inadequately expressed by neoclassical economists associated with the MPS as the proposition that prices in an efficient market "contain all relevant information" and therefore cannot be predicted by mere mortals, whose powers fail to measure up. Whatever the version, the moral is always the same: the market always surpasses the state's ability to process information, and this constitutes the kernel of the argument for the necessary failure of socialism.

2. *Neoliberalism thoroughly revises what it means to be a human person.* Many quote Michel Foucault's prescient observation from over three decades ago: "In neoliberalism ... Homo Economicus is an entrepreneur, an entrepreneur of himself."[7] However, they tend to overlook the extent to which this is a drastic departure from classical liberal doctrine.

Classical liberalism identified "labor" as the critical original human infusion that both created and justified private property. Foucault correctly identifies the concept of "human capital" as the signal neoliberal departure from classical liberal thought—initially identified with MPS member Gary Becker—that undermines centuries of political thought that parlayed humanism into stories of natural rights. Not only does neoliberalism deconstruct any special status for human labor to ground the legitimacy of property rights, but it also lays waste to older distinctions between production and consumption that were rooted in the labor theory of value, and reduces the human being to an arbitrary bundle of "investments," skill sets, temporary alliances (family, sex, race), and fungible body parts. "Government of the self" becomes the taproot of all social order. And more to the point, "knowledge" as a cognitive state becomes hopelessly conflated with knowledge as a thinglike commodity in human capital theory.

3. Most neoliberals insist that they value "freedom" above all else; but more hairs are split over the definition of freedom than over any other neoliberal concept. Some members of the MPS, like Milton Friedman, refuse to define it altogether (other than to divorce it from democracy), while others, like Friedrich Hayek, proceed by motivating it as an epistemic virtue: "the chief aim of freedom is to provide both the opportunity and the inducement to insure the maximum use of knowledge that an individual can accrue."[8]

As this curious definition illustrates, for neoliberals, what you think a market really is seems to determine your view of what liberty means. Almost immediately, the devil is secreted in the details, since Hayek feels he must distinguish "personal liberty" from subjective freedom, since personal liberty does not entail political liberty.[9] Nevertheless, it demonstrates the absolute pivotal character of "knowledge" in their political economy.

It is axiomatic that "freedom" can only be "negative" for neoliberals (in the sense of Isaiah Berlin), for one very important reason. Freedom cannot be extended from the use of knowledge *in* society to the use of knowledge *about* society, because self-examination concerning why one passively accepts local and incomplete knowledge leads to contemplation of how market signals create some forms of knowledge and squelch others. Meditation upon our limitations imposed by dependence upon markets leads to inquiry into how markets actually work, as well as meta-reflection on our place in larger orders, something that neoliberals warn is beyond our ken. Knowledge in that eventuality assumes global dimensions and is no longer "local," and this undermines the key doctrine of The Market as a transcendental superior information processor. Conveniently, "freedom" does not extend to principled rejection of the neoliberal insurgency.

4. Neoliberals regard inequality of economic resources and political rights, not as an unfortunate by-product of capitalism, but as a necessary functional characteristic of their ideal market system. Inequality is not only the natural state of market economies from a neoliberal perspective, but it is actually one of its strongest motor forces for progress. Hence, the rich are not parasites but, rather, a boon to humankind. People should be encouraged to envy and emulate the rich. Demands for equality are merely the sour grapes of the losers, or if they are more generous, the atavistic holdovers of older images of justice that must be extirpated from the modern mindset. As Hayek wrote, "the market order does not bring about any close correspondence between subjective merit or individual needs and rewards."[10]

Indeed, this lack of correlation between reward and personal effort is one of the major inciters of (misguided) demands for justice on the part of hoi polloi. "Social justice" is blind, because mere humans can never comprehend the true consequences of their market activity. Inequality in resources is paired with ineradicable inequality in knowledge: in a well-functioning economy, most people are simply doomed to be mired in stupidity.

5. The Market (suitably reengineered and promoted) can always provide solutions to problems seemingly caused by the market in the first place. This is the ultimate destination of the constructivist political program within neoliberalism. Any problem, economic or otherwise, has a market solution, given sufficient ingenuity: pollution is abated by the trading of "emissions permits"; inadequate public education is rectified by "vouchers"; auctions can adequately structure exclusionary communication channels; poverty-stricken sick people lacking access to health care can be incentivized to serve as guinea pigs for privatized clinical drug trials; poverty in underdeveloped nations can be ameliorated by "micro-loans."

Those closer to the neoclassical orthodoxy were quick to take up the entrepreneurial mantle: suitably engineered boutique markets were touted as a superior method of solving all sorts of problems previously thought to be better organized by governments—everything from scheduling space shots to regulating the flow through airports and national parks. Market design is the panacea for whatever ails you. Economists made money selling their nominal expertise in setting up these new custom-engineered markets, rarely admitting upfront that they were simply acting as middlemen introducing intermediate steps toward subsequent full privatization of the entity in question.

6. *The neoliberal project has thoroughly revised law; but especially, criminal law.* Members of the Mont Pèlerin Society were fond of Benjamin Constant's adage: "The government, beyond its proper sphere ought not to have any power; within its sphere, it cannot have enough of it."[11] It is central to understanding the fact that neoliberal policies lead to unchecked expansion of the penal sector, as has happened in the United States. As Bernard Harcourt (2011) has explained in detail, however much some neoliberals might seem to suggest that crime be treated as just another market process, neoliberals have moved from the treatment of crime as exogenously defined within a society by its historical evolution, to a definition of crime as inefficient attempts to circumvent The Market. The implication is that intensified state power in the police sphere (and a huge expansion of prisoners incarcerated) is fully complementary to the neoliberal conception of freedom. In the opinion of MPS member Richard Posner, "The function of criminal sanction in a capitalist market economy, then, is to prevent individuals from bypassing the efficient market."[12]

In this neoliberal perspective, there is also a natural stratification in what classes of law are applicable to different scofflaws: "the criminal law is designed primarily for the nonaffluent; the affluent are kept in line, for the most part, by tort law."[13] In other words, economic competition imposes natural order on the rich, because they have so much to lose when it is violated. The poor need to feel the naked violence of the law.

This short doctrinal gloss constitutes little more than a Cliff's Notes caricature, intended to evoke the culmination of decades of work by the neoliberal movement, beginning with the first meeting of the MPS in 1947. We have no doubt some modern neoliberals will find it excessively curt and schematic, if not wrong-headed. However, our task here is not to faithfully describe the subtleties of the many tenets of modern neoliberalism, nor to explore the ways its lineage does or does not terminate in a single monolithic creed.[14] It is, rather, to provide a glimpse of the indispensable role of the MPS neoliberals in shifting the fundamental conversation about markets away from the earlier neoclassical mantra of "the allocation of scarce resources to given ends" and ever closer toward the The Market as Super Information Processor. This will be rendered more plausible after we sketch the historical steps by which this occurred.

[5]
THE SOCIALIST CALCULATION CONTROVERSY AS THE STARTING POINT OF THE ECONOMICS OF INFORMATION

Confining ourselves in this chapter to matters of internal intellectual developments *within* the economics profession, we note that the most relevant set of events for the birth of the modern economics of information was a dispute originating within the Austrian tradition—namely, the Socialist Calculation Controversy.[1] The starting gun was a provocative paper by Ludwig von Mises, insisting socialism had to fail because dispensing with market valuations would render all rational calculation impossible. Beginning in the 1920s, a stellar array of economists (and even some philosophers, such as Otto Neurath) began to argue either for or against the proposition that socialist planning was impossible in principle, owing primarily to epistemic considerations. While we cannot retrace every twist and turn in that argument here, there are a few salient points it will prove useful to remember when it comes to discussing the economics of knowledge.

The Socialist Calculation Controversy has sometimes been dramatized as a battle between Old World Vienna and the New World neoclassicals at the Cowles Commission in the United States.

While that plot line is a bit abridged, it will serve for our current purposes. For an important phase of its life span—roughly from 1938 to 1954—the Cowles Commission was located at the University of Chicago, an inconvenient fact that sometimes creates problems for people nowadays who use the term "Chicago" to refer to a distinct reactionary brand of political economy. In the 1940s and 1950s, "Chicago" might refer to the neoliberal circle around Aaron Director and Milton Friedman, or it might equally refer to Cowles's high-powered mathematical economics. The Mont Pèlerin Society (MPS), founded in 1947, was ground zero in the development of neoliberalism, and it enjoyed a significant beachhead at the University of Chicago after WWII, owing to the efforts of Milton Friedman, Aaron Director, and Allen Wallis.[2] Clearly, Chicago was an intense and lively place to be around 1950.

The first point necessary to stress is that the Cowles Commission soon became the rallying ground for those who sought to rebut and refute Vienna, and later the MPS, with regard to the asserted unintelligibility of rational thought under socialism. Jacob Marschak, who became a research director of the Cowles in 1943, published his first article back in Vienna, in 1923, as a response to Mises in the calculation controversy. His predecessor as research leader of Cowles in the later 1930s, Oskar Lange, published *On the Economic Theory of Socialism* (originally 1936) to upbraid Mises with mathematics and neoclassical price theory. Leonid Hurwicz, a central character in our drama, actually studied with Mises; his later work at the Cowles on mechanism design was promoted explicitly to refute him.

As early as 1935, Mises's protégé Friedrich Hayek mounted a counterattack on all those neoclassical market socialists by slightly shifting the terms of debate—and here we find the flapping wings of the butterfly that would result in a hurricane of chaotic economics of information late in the century. There was not a single chaotic trajectory; rather, there were at least three different paths of orthodoxy, as we shall argue in chapter 8.

The best, quick place to start any brief account of the Socialist Calculation Controversy is with Friedrich Hayek. Today, Hayek is quite famous, familiar to viewers of YouTube, denizens of Silicon Valley, and members of the Tea Party; but that does not mean these modern admirers understand his subtle ideas concerning knowledge and information, and how they changed over his lifetime.

The cardinal insight of the early Hayek was to abandon Mises's strange insistence that all "calculation" whatsoever would be impossible under socialism, and replace it with the seemingly more credible proposition that it would be impossible to collate and deploy all the knowledge required to coordinate the economy as successfully as the market managed to do in practice. In other words, he transformed what Mises had portrayed as a breakdown of (Max) Weberian "Zweckrationality" under socialism into something initially far less threatening, a species of epistemological difficulty endemic under socialism.[3] The error of socialism, said Hayek, was to try and accomplish something through planning that had already been solved by The Market. His proposal to change the subject of economics was made compellingly in his 1945 essay "The Use of Knowledge in Society":

> What is the problem we try to solve when we try to construct a rational economic order? On certain familiar assumptions the answer is simple enough. *If* we possess all the relevant information, *if* we can start out from a given system of preferences, and *if* we command complete knowledge of available means, the problem which remains is purely one of logic. . . . This, however, is emphatically *not* the economic problem which society faces. . . . The peculiar character of the problem of a rational economic order is determined precisely by the fact that the

knowledge of the circumstances of which we must make use never exists in concentrated or integrated form but solely as dispersed bits of incomplete and frequently contradictory knowledge which all the separate individuals possess. The economic problem of society is thus not merely a problem how to allocate "given" resources . . . it is a problem of the utilization of knowledge which is not given to anyone in its totality.[4]

Here we witness the birth of the First Commandment of neoliberalism. Markets don't exist to allocate given physical resources, so much as they serve to integrate and disseminate something called "knowledge." The Market had ceased looking like a mechanical conveyor belt, only to take on the vague outlines of the computer. What is striking in retrospect is not so much that this new definition of markets swept the MPS; rather, it is the extent to which market socialists, especially the ones at the Cowles—the intended targets of Hayek, rapidly took it on board. Herbert Simon favorably quoted the passage.[5] Paul Samuelson claimed in retrospect it was the starting gun of "information economics."[6] Even Bank of Sweden Prize winner Robert Solow gave it his imprimatur:

The Good Hayek was a serious scholar who was particularly interested in the role of knowledge in the economy. . . . All economists know that a system of competitive markets is a remarkably efficient way to aggregate all that knowledge while preserving decentralization.[7]

The fact that this new image of markets as superior information processors so comprehensively swept everyone along—neoclassical theorists, market socialists, and neoliberals—with almost no serious scrutiny or skepticism is one of the more astounding facts of the latter twentieth century in economics. All you had to do was read a little Hayek to realize he was rejecting the basic approach of neoclassical microeconomics from the 1940s onward; but that didn't seem to bother many orthodox economists. A tiny virus of meaning injected into economics in the 1940s so thoroughly took over that markets became an utterly different species by the 1990s, if not sooner.

If we return to our Nobel Prize listing, table 4.1, we can now see how the order of the prizes slightly misrepresents the chronology of the economics of information. It was Hayek and the Socialist Calculation Controversy that came first; then, most of the other Bank of Sweden Prize winners responded to this innovation in their own idioms and peccadillos, be they the language of MPS or the American neoclassical orthodoxy. As some Austrian economists have observed, "information economics is the most prominent example of how formal theorists have attempted to translate Hayek's ideas into a form easily digested and incorporated by mainstream economists."[8] Some, like Samuelson and Arrow, got their Swedish awards first, but in fact, they were reacting to the preeceding Socialist Calculation Controversy.

Whatever their special innovation, few if any of these enablers bothered to expound upon the precise nature of this "knowledge" that the market so effortlessly and efficiently processed (as we have already witnessed with Arrow), much less bothered to find out what Hayek had actually meant by the term. The next generation brashly presumed their own personal, idiosyncratic meaning of "knowledge" would be abundantly transparent to everyone else, and in many instances, could be easily grafted onto neoclassical microeconomics. What a lovely paradox: every economist simply presumed that everyone else knew what they were talking about when they sang the praises of the magic capacities of the market to convey knowledge; but if it were always that easy to get ideas across, in such effortless mental harmony, then who needed The Market in the first place?

The place of information in economics was broached in heated disputes over the politics and possibilities of socialism. The fascinating thing about this fact is how it all tended to get neglected, soft-focused, and even forgotten over the decades, so that theoretical disputes might appear as low-temperature technocratic propositions instead. The student should realize, though, that the original context always lurked in the background, even generations later.[9]

The remainder of this book aims to describe how this extraordinary train of events unfolded. We start by recounting Hayek's own struggles with the knowledge concept, then suggest that Hayek helped constitute the space of "thinkable information" for both the Austrians and the orthodox neoclassical economists. We then subsequently turn to the more familiar orthodox figures, like Arrow and Hurwicz and Stiglitz, and reveal how their problem situation was shaped by Hayek.

[6]
HAYEK CHANGES HIS MIND

Only recently, with the explosion of historical literature on Hayek, have we begun to encounter serious scholarly work on Hayek's struggles with epistemology.[1] As with almost every other major intellectual figure, Hayek changed his position on key theoretical terms over the course of his career; and none was more consequential than his treatment of knowledge. Interestingly, in Hayek's last book, *The Fatal Conceit*, he admits, "I confess it took me too a long time from my first breakthrough, in my essay "Economics and Knowledge" through the recognition of "Competition as a Discovery Procedure" and my essay on "The Pretense of Knowledge" to state my theory of the dispersal of information, from which follows my conclusions about the superiority of spontaneous formations to central direction."[2] So while we have his frank admission that his system did not congeal around the concept of *information* until rather late in his career, at least in his own mind, we do not have a corresponding historical schematic of how it changed in his own writing. Leaning on the secondary literature, we will proceed to summarize it as a concerto in three movements.

We have already been introduced to the first movement of this concerto. Hayek displaced the rather cryptic position of Ludwig von Mises in the Socialist Calculation Controversy with the notion that knowledge is "dispersed" in such a way that bringing it all together in a central planning authority would be difficult—but, note well, *not impossible*. There seemed to be a special kind of slippery knowledge, a glistening goo qualitatively different from more conventional scientific conceptions, that was *local*, characterized by special conditions of time and place. It was almost as if this species of knowledge was something *entropic*: an energy that grew too diffuse to be readily gathered up and consolidated into a useful form.[3]

Not all knowledge shared this character, said Hayek; but the mere fact it existed at all gave him a club with which he could beat up the Langes and Marschaks of this world. Sometimes, Hayek hinted that the dispersed character had something to do with subjective experience, but at this stage he steered well clear of issues of cognitive capacities or capacities to articulate this knowledge to others. In this movement, there was very little in the way of actual epistemology or formal psychology standing behind the concept. Instead, in his famous paper "Use of Knowledge in Society" (1945), he proposes to reconceive the market as a "mechanism for the communication of information." Perhaps this is one reason it seemed to appeal to some neoclassical economists, who were more readily inclined to interpret knowledge of this ilk as a "thing" scattered about the landscape, rather like pixie dust too fine to pick up. Indeed, most of the favorable citations of Hayek by neoclassical economists date from this period.

The next movement in Hayek's Surprise Concerto happened some time around his own return to psychology, published in 1952 as *The Sensory Order*. As he tells us in the preface, that book was based on a student paper he had written thirty years previous, which "contained the whole principle of the theory I am now putting forward."[4] At this stage, Hayek entertained the notion that much of human knowledge is not only inarticulable but also tacit and inaccessible to self-examination. As he complained, "It has undoubtedly been unfortunate for the development of psychology that the distinguishing attribute of its object was long to be considered the "conscious" character of experience,"[5] something he believed he could counteract.

Much of his revised attitudes concerning knowledge seem to have occurred during his stint at the Committee on Social Thought at the University of Chicago. In brief, Hayek there sought to revive the old, discredited associationist psychology of the late eighteenth and early nineteenth century, by suggesting that the mind was little more than sets of hierarchies of systems of classifier algorithms, which were opaque to the thinker.[6] He also had been in contact with Michael Polanyi at the early MPS meetings, and had come across Gilbert Ryle's distinction between "knowing how" and "knowing that" in Ryle's *Concept of Mind* (1949). He began to explore variations on "tacit" or nonarticulable knowledge, not so much by explicitly following Polanyi or Ryle on this topic as by building his own idiosyncratic theory of mind upon a foundation of classifier systems about which the subject was not even aware, but regularly made use of in order to interact with the environment.[7]

From this point forward, Hayek began to play fast and loose with the concept of consciousness, inverting the then-popular Freudian frame tale that the unconscious was a soup of barely accessible urges upon which rested a fragile vessel of rational thought; for Hayek, it was *rationality that was largely unconscious,* with conscious perception and drives constituting the veneer of intentionality and desires floating on top of the sea of obscure and inaccessible rule structures. Thus, the types of knowledge that mattered most were inarticulate and largely inaccessible to the thinking agent.

It was also precisely at this juncture that Hayek began making explicit references to evolutionary theory as the basis of his entire philosophy. The reason behind this shift was that Hayek sought to propound that the individual mind did not actually choose the rules that worked the best; that was done either through a sort of quasi-evolutionary selection of life success at the individual level reinforcing the relevant classifier rules, or more frequently, through natural selection's weeding out the individuals with unfit rules in favor of those individuals lucky enough to come previously equipped with superior classifiers. It was, not to mince words, a harsh version of social Darwinism.

It is important to understand how this refracted the very notion of radical ignorance as a natural state of being for humankind in the later political economy of Hayek.[8] In this conception, the process of coming-to-know became largely disengaged from the knower, with most of the action happening at the subconscious level. As he wrote in his "Primacy of the Abstract," "the formation of a new abstraction seems *never* to be the outcome of a conscious process, not something at which the mind can deliberately aim, but always a discovery of something which *already* guides its operation."[9] Here, the celebrated philosopher of freedom postulated a grim species of predestination that would make even Calvin blush. The political implication was clear: if an individual mind could not even reliably plan or organize its own pathway of learning through life, it would exhibit contemptible hubris to think it could ever plan the lives of others, much less a whole economy. Knowledge here was no longer like entropy or pixie dust; now it resembled a great submerged iceberg, nine-tenths of it invisible, and all of it frozen into place eons ago, with only minor changes around the margins when it jostled up against other icebergs.

How did these lumbering monads ever manage to communicate, much less live in societies that displayed any reliable level of organization? That question was finally answered in the third movement of Hayek's Surprise Concerto. Strangely for a doctrine that started out so concerned about respect for the inviolate individual and his or her subjectivity, the late Hayek rendered his system internally coherent by admitting that some knowledge did not really persist at the level of the individual mind, for the most part, but was processed and invested with meaning at the supra-personal level. In a catch phrase, since so much that people actually knew was inaccessible to them, the only entity that really was capable of judging and validating human knowledge was The Market. The key turning point, as Hayek informs us in *The Fatal Conceit*, was his essay "Competition as a Discovery Procedure" (1968):

> [Epistemology is governed by] competition as a procedure for the discovery of such facts as, without resort to it, *would not be known to anyone*. . . The knowledge of which we speak consists rather of a capacity to find out the particular circumstances, which becomes effective *only if the possessors of this knowledge are informed by the market* which kinds of things or services are wanted, and how urgently they are wanted. . . . Knowledge that is used [in a market] is that of *all its members*. Ends that it serves are the separate ends of all those individuals, in all their variety and contrariness.[10]

No longer was knowledge being treated as an elusive thing by Hayek, scattered about in an inconvenient matter; in this version, not only is much human knowledge unable to be retrieved from within by the individual in question but, indeed, there exists a species of *knowledge not "known" by any individual human being at all*. Here we are cosseted in the realm of Donald Rumsfeld's infamous "unknown unknowns."[11]

Now what is the message here? The message is that there are no "knowns." There are things we know that we know. There are known unknowns. That is to say, there are things that we now know we don't know. But there are also unknown unknowns. There are things we don't know we don't know. So, when we do the best we can and we pull all this information together, and we then say "Well, that's basically what we see as the situation," that is really only the known knowns and the known unknowns.

The only recourse of the rational individual in this subpar situation is primarily to acquiesce to the dictates of signals conveyed by The Market, which hint at deeper truths than most humans will ever know.

But what is this depersonalized and deracinated supra-human knowledge but a new virtual kind of *information*? This, we think, explains Hayek's rather uncharacteristic reversion to replacing the term "knowledge" with "information" in his last work, *Fatal Conceit* (1988). There, he wrote:

> Comprehending the role played by the transmission of information (or of factual knowledge) opens the door to understanding the extended order. Yet these issues are highly abstract, and are particularly hard to grasp for those schooled in the mechanistic, scientistic, constructivist canons of rationality that dominate our educational systems.[12]

Sometimes, when it came to this ectoplasmic information, the late Hayek lapsed into his scientist mode, where evolution had winnowed the elusive truth out of human frailty; but other times, he reverted to full religious mystery: "spontaneous order ... cannot be properly said to have a purpose ... known to any single person, or relatively small group of persons."[13]

Some latter-day Austrians have argued that entrepreneurs are just "smarter" than any dedicated intellectual, since they are marinated in this information and thus quicker to respond to market signals.[14] Yet, almost by definition, there is no instrument available to humankind to "test" this proposition. As with all the great world religions, the sole and final terminus for the skeptic is to surrender to faith: The Market as a super information processor knows more than we could ever begin to divine.

One might aver that this is an egregiously idiosyncratic trajectory, the ruinous road to the conflation of pervasive ignorance with virtue, something that would never, ever be followed by any prudent rational choice orthodoxy in economics nor, indeed, any scientific thinker.[15] The modern economist often claims to like the early Hayek, but thinks he studiously avoids the later Hayek. In this book, we beg to disagree: the historical record is far richer than that. Indeed, what we will refer to from now on as the "Hayek trajectory" constitutes one axis of the conceptual space of "information thinking" that we shall proceed to put to use in organizing the subsequent history of *orthodox information economics*. On this axis, "information" goes from being merely difficult to retrieve, to being partly inaccessible, to becoming finally so transcendent no one can really know it. While Hayek and the neoliberals were primarily responsible for this epistemology axis, it was the *nouvelle vague* of neoclassical microeconomists at Cowles that was primarily responsible for the technical axis of multiple mathematical instantiations of information, the subject of the next two chapters.

[7]

THE NEOCLASSICAL ECONOMICS OF INFORMATION WAS INCUBATED AT COWLES

We can imagine many readers who are not of an Austrian persuasion beginning to wonder just where this particular story is headed, and more important, whether it is going to confront what "most people" think is the real, practical content of modern orthodox economics of information, such as asymmetric information, game theory, incentive compatibility, and so forth. Rest assured; we have finally arrived at the juncture where the hallmark doctrines begin to make their appearance. All the standard protagonists are here—Arrow, Marschak, Hurwicz, and others—yet Hayek turns out to be influential in this story because of a curious quirk in the history: the standout group of neoclassical high theorists who successfully established Walrasian general equilibrium models as the Ten Commandments of the nascent postwar American orthodoxy were self-professed market socialists—at least in the 1940s and early 1950s.

There is, as yet, no really comprehensive history of the role of the Cowles Commission in the development of economics in the United States.[1] The Cowles Commission was chartered in Colorado Springs in 1932 by the businessman Alfred Cowles, initially to serve as a sort of boutique research unit to argue against the Depression-era National Recovery Administration (NRA), to pursue monetary reforms, and to explore what went wrong with the stock market. When Alfred Cowles himself was forced to move to Chicago in 1939, he worked out a deal with University of Chicago's President Robert Hutchins and trustee Laird Bell to have the unit enjoy a semi-formal relationship with the university.[2] Economics faculty members Oscar Lange, Jacob Mosak, and Gregg Lewis were given positions as part-time staff, which began to change the tenor of work being done there. It was only at this juncture that Cowles became narrowly neoclassical in theoretical orientation, although this stance was not yet common in the United States. World War II had severely depleted the staff at Cowles, and by the mid-1940s, its continued existence was in doubt. A new round of hiring, beginning with Leonid Hurwicz in 1944 (after serving as an assistant to Lange and Theodore Yntema), and Jacob Marschak in 1943, began to recast the unit in its more recognizably modern guise, and in turn, staff the commission with new economics faculty. That novel identity was cemented in place with the appointment of Tjalling Koopmans as research director in 1948 (having been appointed associate professor at Chicago in 1946), and the forging of direct ties with RAND and the military thenceforth.[3] Kenneth Arrow was a research associate in 1947, and was appointed to the faculty in 1948, at which point Cowles was permitted to propose faculty hires solely from within the unit.

With hindsight, we now can appreciate that the Cowles Commission was the citadel of this political movement to forge a market socialism—at least until that organization picked up and moved to Yale in 1955. This position had at least two signal implications.

First, Cowles members jointly and severally felt they had to respond to, and possibly refute, Hayek and the MPS representation at the Chicago economics department, because the Socialist

Calculation Controversy was the greatest intellectual threat to their vision for the role of the scientific economist in society. Furthermore, stoking their motivation from 1950 onwards, Hayek was physically there, on the ground at Chicago, lodged in the Committee on Social Thought. In essence, Austria had come to Chicago, to the consternation of all and sundry. Marschak and Hurwicz had already clashed with Hayek and comrades well before this galling development; now they inhabited the very same campus. As early as 1940, Hurwicz found the economists at Chicago "very reactionary and orthodox. I met Viner, Knight and the other local celebrities . . . and didn't think very much of them."[4] Marschak had been one of the readers for the University of Chicago Press of Hayek's manuscript of *Road to Serfdom*; he did recommend publication, but made clear his disagreements with its contents, noting "the book is almost exclusively critical, not constructive."[5]

Marschak opposed creating a Frank Knight chaired professorship at Chicago, on the following grounds: "it is probably unavoidable that in filling a Frank H. Knight chair preference would have to be given—always, or at least for the next twenty years—to followers of a particular orientation in economic policy, even when candidates of higher scientific objective merit were available. Would it be possible to honor the great Martin Luther's memory by a chair and offer it to an outstanding Catholic thinker?"[6] Others at Cowles were no less stalwart in their commitment to refighting the Socialist Calculation Controversy after World War II. Tjalling Koopmans prefaced the proceedings of the Cowles conference on "linear programming" with the following observation:

> Particular use is made of those discussions in welfare economics (opened by a challenge of L. von Mises) that dealt with the possibility of economics calculation in a socialist society. The notion of prices as constituting the information that should

circulate between centers of decision to make consistent allocation possible emerged from the discussion by Lange, Lerner and others.[7]

There is plenty of evidence the Cowlesmen (many of whom were European exiles) were feeling a little embattled at Chicago, especially after the war. Even Arrow, the home-grown Cowlesman has testified to his political motivations at that juncture:

> On returning from military service, I planned to write a dissertation which would redo *Value and Capital* properly, a very foolish idea. I had two motivations. One was to supply a theoretical model as a basis for econometric estimation. The other was a strong interest in planning. I would have described myself as a socialist, although one that had a strong belief in the usefulness of markets. Market socialism was a widespread view. Hotelling held it. It had been popularized especially by the works of O. Lange.[8]

Second, the Cowlesmen believed they would accomplish this political task *not* by following Hayek and the others into the thickets of some school or other of formal psychology, but instead, by producing their own, purpose-built version of the mathematical utilitarian mind, developing a novel "decision theory" based upon *avant garde* natural science currents contemporary with their efforts. Much of what passes for modern decision theory, therefore, had its origins not in economics per se but, rather, in operations research, growing out of WWII.

Recalling the insight of Hunter Heyck, the military was inclined to shift the focus from the mental state of the chooser to the *choice* as a freestanding phenomenon worthy of study; operations research was the vessel that supported the reorientation.[9] Because

Cowles developed a close and lasting research cooperation with the military at RAND starting in 1948, it was specifically the Cowles branch of American neoclassicism that introduced operations research into economics, and later, into business schools across the nation.[10] The history of the military inspiration of "rational choice" cited in chapter 1 documents this fateful nexus.

Thus, we observe that it was heavily over-determined that it would be Cowles economists who were initially recruited to the front line, dedicated to confronting the nature of "information" in economics, not because they were especially subtle specialists in epistemology but, rather, because their patrons promoted it, their politics dictated it, and everything about their own commitments as to what a scientific economics required demanded they take a position on it. This even extended to their defense against the most withering criticisms of neoclassical theory in the American context, most of which had emanated, not from Austrians, but from the American Institutionalists.

The Cowlesmen had decided by the late 1940s that the Institutionalists were their immediate sworn enemies—to an even greater degree than the Marxists. In a catch phrase, those obstreperous Institutionalists had accused the earlier neoclassicals of tying themselves in cognitive knots. In the early twentieth century, went the accusation, neoclassical theory had reached a strange impasse.[11] By early century, neoclassical economists wanted to renounce any dependence upon psychology, rejecting the guise of "utility" as a mental phenomenon. But simultaneously, they sought to address the common complaint that their theory was epistemically implausible—that it was a theory simply assuming "perfect knowledge" of desires, prices, and possibilities, in order to place "knowledge" out of bounds of the theory. How could there be such a "brainless/mindless" type of knowledge, built up around noncognitive "preferences"? How could such passive brain-dead zombie agents "know" that the market gave people what they really wanted? This was the conundrum that launched a thousand leaky vessels at Cowles.

The lead vessel in the armada contained a rather simplistic notion: treating information as a "thing" subject to market organization initially seemed to offer a blissful resolution beyond the impasse. Such an object-oriented construct was only the *first* draft of a formal incorporation of information into economic theory. But in conjuring information as fugible object, neoclassical theory took another vertiginous turn at Cowles, one which we can only gesture toward in this narrative.

It has now become commonplace among historians of economics to realize that before the early twentieth century, both classical and neoclassical economics were *past-oriented*. That is, the values realized today were explained as the consequences of events that had happened yesterday, in the past: labor values were due to past infusions of labor; Marshallian supply prices were the consequence of prior production and investment decisions; "final degree of utility" was the consequence of prior consumption choices. From Frank Knight's *Risk, Uncertainty and Profit* ([1921] 1964), price theory swiveled 180 degrees on the time axis, and economic theory rather quickly instead became *future-oriented*. One way to summarize this curious transformation is that it subsequently became commonplace to assert that *events which had not yet happened* could come to influence economic decisions in the present, in a kind of spooky action-at-a- distance. Past decisions, by contrast, were treated as irrelevant bygones, "sunk costs" no longer meriting consideration.

The introduction of inductive statistics from the 1930s forward reinforced this dramatic sea change, in the sense that current values would henceforth be said to embody an irreducible component of prospective future risk. Keynes claimed that much of the breakdowns of the 1930s were due to irreducible uncertainty about the future. Once probability theory was married to utility theory in the 1940s—a specialty at Cowles—their "knowledge" became knowledge about the future consequences of current decisions, and it fed back directly into those decisions. If the inscrutable future was conceived to cause economic changes in the present, then "information" became the number one causal cue though which this happened.

The conception of information as embodied in a technology of inductive inference was the *second* version of a formalized economics of information. A *third* draft of the economics of information grew out of experience with the early digital computer at RAND and Cowles. All of these components came together at the postwar Cowles Commission.

THE DRAMATIS PERSONÆ

The advent of information processing at Cowles was the confluence of a far more complex set of events than the spare vignettes recounted here, and cannot be done justice in a few paragraphs.[12] Nevertheless, a crude gloss would point to the fact that Jacob Marschak, Kenneth Arrow, Herbert Simon, Stanley Reiter, and Leonid Hurwicz (among others) were all heavily influenced by contemporary developments within cybernetics (see table 7.1); but that they all started out (under the influence of Claude Shannon and RAND) by treating information as a fungible commodity— "Uncertainty usually creates a still more subtle problem in resource allocation; information becomes a commodity,"[13]—only to rather rapidly back off from this option (although never entirely denouncing it), then to subsequently transfer allegiance to conflating information processing with statistical induction.

This move was something they learned by watching the operations researchers; but it was also closely related to their retreat from

Table 7.1 COWLES COMMISSION MEMBERS AND THEIR INFORMATION ENTHUSIASMS

Cowlesmen	Period at Cowles	Information Innovations
Jacob Marschak	1943–1960	Economics of information, team theory, early experimental economics
Leonid Hurwicz	1942–1951	Incentive compatibility, mechanism design
Kenneth Arrow	1947–1950	Moral hazard, many papers on information
Stanley Reiter	1948–1950	Mechanism design, computability
Herbert Simon	1947–1949	Artificial intelligence, bounded rationality, computation

a full-blown econometric empiricism (for which they had originally gained recognition) in favor of models of the economic agent as a miniature econometrician. At that point, the original Cowles team spun off into various semi-unrelated research programs into the economics of information, as the allure of purely epistemic econometrics palled. Eventually, however, Cowles members in various combinations would conduct reconnaissance into all three technical paradigms of information analysis, as enumerated in the next chapter.

One should not glean a mistaken impression of fleeting attention from the limited span of time some of the figures listed in table 7.1 spent at Cowles; their sojourn in Chicago was decisive for each of them when it comes to information economics. Let us very briefly explore this landscape as it developed in the critical period 1948 to 1954. We shall begin with Tjalling Koopmans, who does not really make our league table, but helped set the stage nonetheless.

Tjalling Koopmans

Some members of Cowles started out believing that the existing Walrasian model, as formalized by Arrow and Debreu, was sufficient in and of itself for refutation of Hayek's proposed revision of the marketplace of ideas. Tjalling Koopmans adopted the position that the Walrasian model then being reformulated at Cowles by Arrow and others actually showed that agent cognition was effectively unnecessary, since the individual agent only needed to know his own preferences and parametric prices in order for equilibrium to obtain:

> [O]ne can in particular interpret the proposition as a statement of conditions under which the simplicity of incentive structure and the economies of information handling characteristic of a competitive market organization can be secured without loss of efficiency of allocation. . . . The price system carries to each producer, resource holder, or consumer a summary of information about the production possibilities, resource availabilities and preferences of all other decision makers. Under the conditions postulated, this summary is all that is needed to keep all decision makers reconciled with a Pareto optimal state once it has been established.[14]

Perhaps the most incongruous aspect of this assertion was its utter lack of connection to the Cowles stated political orientation of market socialism, or its justification. Somehow arriving at this amazing state of coordination dispelled any need for communication or thought. Koopmans's endorsement of the amazing powers of general equilibrium came a little too close to sounding like Hayek, which may explain why the other Cowlesmen so noticeably distanced themselves from this brutally limited construal of "information."

Other members of Cowles were not quite so publicly confident that their sanctioned heritage of Walrasian models adequately addressed this supposed exquisite effortless economy of information. Marschak and Hurwicz in particular were more inclined toward doubt, and Arrow (as usual) began to cast about for various glitches that frustrated the welfare theorems. Much discussion within Cowles initially sought to bundle together the various worries and qualms onto the Procrustean bed of uncertainty, and one can observe by the mid-1950s Koopmans first floated the trial balloon of blaming this insufficiency on "missing markets" (later popularized by Arrow):

> Here, perhaps the most crucial kind of uncertainty . . . arises from the lack of information on the part of any one decision-maker as to what other decision-makers are doing or deciding to do. It is a puzzling question why there are not more markets for future delivery through which the relevant information about concurrent decisions could circulate in an anonymous manner.[15]

Jacob Marshak

Koopmans was nowhere quite so daunted by these problems as Jacob Marschak, about whom "on many occasions during the 1950s and 1960s we heard economists question whether Marschak had not actually left economics for other disciplines, such as psychology [or] information science."[16] Yet, far from being a flighty dilettante or fickle fellow traveler, Jacob Marschak was situated at the very core of the Cowles project in the 1950s. For someone whose allegiance to the Walrasian orthodoxy was never in doubt, even to the point of rejecting Keynesian economics,[17] Marschak seemed painfully sensitive to the ways in which "information" might disrupt economic equilibrium. Because he had participated in some early decision theory experiments at RAND and elsewhere, perhaps to a greater degree than his comrades he appreciated the latent empirical failures of decision theory and its offshoot, the theory of expected utility. He early on adopted a stance that has later become second nature in the rise of "behavioral economics": "If we know what makes people more or less logical or mathematically inept or poor decision makers, we may also find out how best to enable them to learn the 'recommended' type of behavior . . . The normative and descriptive analyses complete each other."[18] This convenient fusion of the normative and descriptive into a tautology might be instituted by means of elaborations upon the notion of "information," or so Marschak hoped.

Marschak tentatively tried out various paths toward reaching his own grail of an economics of information (and he was one of the earliest American economists to use the term[19]), but none of them seemed to pan out: first, he struggled with subjecting Shannon information to a supply/demand framework, only to reject that; and then subsequently entertained the Blackwell formalism, which suggested an instrument reading was more informative if it could distinguish observations over a finer partition of the state space of possibilities, only later to reject it; later he dallied with the idea of transactions costs as capturing informational issues; he also pioneered a computer/organization metaphor in the format of what he dubbed "team theory."

Marschak taught one of the very first courses anywhere on The Economic Theory of Information and Organization at Yale in the late 1950s, even using Ross Ashby's *Introduction to Cybernetics* as a required text. He corresponded with AI progenitor John McCarthy concerning his paper "Measure of the Value of Information" in 1957.[20] He was among the first to participate with professional psychologists in experiments designed to test the limits of decision theory, back when that was still an anathema within the economics profession.

As late as 1966, he was still mulling over the relative merits of what we shall argue in chapter 8 were two of the three major options for a formal neoclassical theory of information in that era:

> Currently, my primary pet project is "Economics of Information" and in particular the question why communication engineers like to use the entropy formula (presumably needed for purposes of mass producers, not the individual users, of communication equipment) while statisticians and other users must either specify the individual loss function or content themselves with the partial ordering of information systems (also called "experiments") that is induced by Blackwell-Girshick's "greater informativeness" relation.[21]

Marschak's conundrum was that his experimental work had induced him to become rather skeptical when it came to game theory, so he decided he had to concoct a special new field of mathematical economics in order to explore the fine points of information economics; he called it "team theory." As early as 1954, he revealed the centrality of "information" to his concept:

> In a team of executives, each member has to decide something different. These decisions determine an expected joint reward

(payoff) received by the team, and depend on the distribution of information (Who learns what?) among the several partners, and on the decision rules . . . that determine the response of each partner to the information content he receives. The distribution of information depends on the team's communications network and the coding rules used in operating it, and therefore a cost is attached to every form of the distribution of information. The team problem consists in choosing simultaneously a distribution of information and a set of decision rules that yield, in conjunction, the highest expected reward to the team, net of communications costs.[22]

Marschak's "team theory" never really caught on in economics (or anywhere else, for that matter), even though he devoted substantial efforts to its elaboration; but in retrospect, we might entertain it as a somewhat deformed and distended version of what came to be known much later as "mechanism design." But having glimpsed the gleaming shore, Marschak never himself entered the Promised Land.

Leonid Hurwicz

Because the specific development of mechanism design and the path of its progenitor, Leonid Hurwicz, subsequently became so very central to the history of information economics, we will perforce postpone the extended saga of Hurwicz until chapter 9. (A tutorial concerning the mathematics of information is a prerequisite to understand the full story; that is provided in chapter 8.) However, we shall indulge in some preliminaries concerning the influence of Cowles by demonstrating that Hurwicz was already deep into "information processing" at the crucial juncture of the early 1950s, and was discussing it with Marschak, a fact illustrated by the following research outline provided to Marschak, entitled, "Economic Decision-making Processes and their Organizational Structure of Uncertainty":

> The research outlined in the present note is focused on decision-making under uncertainty. The emphasis is, however, not so much on the criteria of optimality among alternative choices as on the *technology* of the processes by whereby decisions are reached and choices are made. Under the customary conditions of "rationality," the final decision is preceded by certain operations which may, in general, be characterized as *information processing* ... when the information processing aspects of the problem are taken explicitly into account it is found that the concept of "rational action" is modified. This is true even when applied to the action of a single individual, but it comes particularly interesting when considered in situations involving many persons. ... The uncertainty need not be generated by external factors like weather prospects: it may be man-made.[23]

The politics of this nascent move seemed somewhat more promising. Hurwicz in the early 1950s was feeling his way toward an account where it was not so much explicit cognitive issues but, rather, the accessory *technologies* that hindered the grand optimum promised by Walrasian general equilibrium.[24] "Market socialism" might then take on a much less threatening coloring of provision of "technological augmentation" of existing markets to achieve full Pareto optima, all in the name of capacities for information processing. Compared to Marschak, it was not so much the "team" that was at fault as it was the accompanying hardware used to convey the price and data messages.

The intention at Cowles back then was still to make use of Walrasian general equilibrium theory in an effort to rebut the political claims of Hayek; yet some bad news on the theoretical front began to queer the market socialist pitch. As a result of the mathematical counterexamples of Herbert Scarf and David Gale in the 1960s, it became clear that the Cowlesmen could give no general dynamic account of global convergence of Walrasian dynamics to a general equilibrium, especially since Hurwicz had been collaborating with Arrow on precisely this question. The legitimacy of full Walrasian equilibrium as the benchmark of ultimate market success would, therefore, seem to have been put at risk. Rather than give up on the Walrasian project, Hurwicz sought a way to "build in" sufficient stability to encourage convergence to an optimum, by conceiving of an economic system loosely as a "convergent computational system." It was, therefore, Hurwicz who first appropriated the Cowles fascination with information and parlayed it into a major subfield of economics, now called "mechanism design."

This set of developments has turned out to be so central to the subsequent history of orthodox information economics that we shall of necessity postpone the details until a later chapter. Yet there is one aspect of Hurwicz's turn that is so idiosyncratic, and yet bearing the unmistakable Cowles stamp, that we must raise it here. Rather than explore a concertedly computational approach to information in the economy, Hurwicz's "mechanism design" studied communication within a tâtonnement-type market system—only now, with the auctioneer gaining the ability to communicate further information than just prices alone. In effect, the commissar was recast as the Great Megaphone. For Hurwicz, the problem with The Market was that its convergence to a Pareto optimum would be thwarted by pesky indivisibilities and nonconvexities; and since these conditions are reputedly not very rare, it should be possible to find another "adjustment process" that could do better. Market socialism (broadly speaking) might offer a way to improve matters, but one could object to such proposals on informational grounds. The point of "mechanism design" was initially to find other adjustment processes that could improve upon performance of The Market without imposing too burdensome a communication requirement.

These studies quickly became wrapped up in considerations of "decentralization," which had concerned Hurwicz since his earlier work on activity analysis (in 1950).[25] Once market socialist proposals underwent redescription as informationally decentralized mechanisms, it became an accepted creed of Cowles market socialists that they, like Hayek, rejected a centralized solution. Socialism thus began to shed its more conventional connotations, and grew distant from any political theory.

But what was the appropriate principle by which to demarcate "centralization" from "decentralization?" Establishing something like this was of fundamental importance to a project that hoped to generate alternative "market-like" non-market allocation mechanisms, where "market-like" was understood to be "decentralized." Hurwicz proposed various definitions of this decentralization, none of which turned out to be especially persuasive, and he was eventually brought to admit, "it is more interesting to see what questions can be asked given a (not *the*) concept of information decentralization."[26]

Presumably it had something to do with the "costs" of transmission, which in practice was conceived as the size or dimensionality of the message space. Eventually, the coterie of mechanism designers around Hurwicz settled on conceiving decentralization as implying a limitation on "channel capacity," and then attempted to ascertain how such limitations restrict the performance of mechanisms. By drawing attention to the importance of restrictions in channel capacity, which was viewed "analogously to a limitation of the (cross-sectional) diameter of a pipe restricting the flow of a fluid through that pipe,"[27] the image before these nascent mechanism designers was, essentially, that of Shannon's information theory. The doctrine that emerged in this tradition was the view that there was a trade-off between performance and information costs, which was deemed the real message of the economics of information.

Herbert Simon

But that did not exhaust the various mutations of "information" at Cowles. The figure of Herbert Simon represents an even more dramatic reprocessing of "information" into whole new disciplinary imperatives, certainly by contrast with Marschak and Hurwicz. The intellectual path of Simon shows us how the various influences and problem situations ricocheting around Cowles could lead in an entirely orthogonal direction, as long as you gave up the severe commitment to maintain fealty to the Walrasian model.

Simon's trajectory is wonderfully covered in his own autobiography,[28] so we provide only a brief summary here. Simon was an odd duck in the Cowles flock: a political scientist specializing in organization theory, but really a polymath. He was invited to sit in at Cowles by William Cooper, while teaching at the Illinois Institute of Technology in Chicago. It would be a challenge to enumerate all the ways he was an outsider to that conclave, but perhaps the most salient was his abiding skepticism toward basic neoclassical microeconomics. In an extremely roundabout manner, this eventually led him to become one of the most important figures in the history of twentieth-century information processing, as one of the three or four founders of the field of artificial intelligence.

Simon did credit his time at Cowles with reorienting his research toward information processing, both in his autobiography and in response to a questionnaire from Clifford Hildreth:

> Perhaps the greatest impact of the Cowles exposure on me was to encourage me to try to mathematize my previous research in organization theory and decision making—especially the theory developed in *Administrative Behavior*. I think this project was on the agenda anyway, but the Cowles contact certainly egged me on and gave it higher priority.... During the period 1950–56, I was doing at least as much economics research as research in management and organizations.... The stint as an almost full-time economist was certainly brought about by my involvement with the Cowles Commission, and later, through the Commission, my association with the RAND corporation. The final unanticipated consequence of these events was to turn me away from economics toward psychology, as my interest in decision making led me to see the need for empirically based theories of human problem solving, and as my RAND consulting brought about my association with Allen Newell and computers.[29]

While most scholarly attention has been devoted to his achievements after his conversion experience with the computer at RAND, our interest here is the underdeveloped narrative of Simon's run-up to that watershed at Cowles, and its relationship to "information." While always believing that the rational choice model was a terrible representation of human thought, one way that Simon fit in was that he, too, was at heart a market socialist. However, he approached his socialism from the side of organization theory and decision making in organizations. Cowles was at the center of all kinds of ferment in mathematical modeling these issues, and this is what attracted Simon to them in the later 1940s. Furthermore, the young Simon harbored ambitions to model human rationality as it functioned in social situations, without much in the way of previous guidance.

One focus of Cowles attention in the late 1940s was game theory, and various mathematical ideas of John von Neumann. Simon was one of the phalanx of Cowlesmen to review *Theory of Games and Economic Behavior* (1944) soon after its appearance; but he was very disappointed in the book as a theory of *organization*—something concerning which he had very strong opinions early on.[30] The axiomatization of expected utility theory found therein also did not appeal. But one thing that did capture his imagination was von Neumann's proselytizing for the newfangled electronic computer, something well represented in Cowles records. Yet, he was further stymied when von Neumann concurrently rejected the computer as a model of the brain, or of psychology in general. The solution to his conundrum came when military work at RAND exposed him to the machine psychology of the man/machine interface at the Systems Research Lab, and consequently he arrived at the position that the scientific objective was to *simulate* human thought, while ignoring basic questions about mind and psychology. This is what it meant to be an information processor: "we began more and more to see decision-making processes as essentially the same as problem solving processes."[31] By his own admission, simulation of heuristics of actual reasoning was good enough to constitute a "theory" of intelligence—hence, the designation "artificial intelligence." Once he arrived at that epiphany, he built a school around him at Carnegie Mellon University that explored man/machine heuristics in greater detail, but also used the template to develop a theory of the corporation patterned on computer metaphors of hierarchy and information processing.

It will be significant for what follows that, although the computer was indispensable to virtually everything that Simon accomplished in social theory after 1955, it was always as a platform for simulation, rather than as a full-blooded application of formal theories of computation to mental phenomena. Simon was intermittently dismissive of those who sought to apply Turing computability to mental processes: "It gradually dawned on computer scientists that the decidability question was not usually the right question to ask about an algorithm or a problem domain."[32]

It is noteworthy that it was renegade Cowles affiliates (such as Alain Lewis, Roy Radner, and Gerald Kramer), and *not* members of the other twentieth-century schools of neoclassical economics, who came to the early realization that tinkering with the utility framework was just too timid a response to the challenge of information, and who struck out to construct a more full-blooded cognitive model. Simon, for instance, kept insisting that rationality was "bounded." Yet it was also plain that the closer the figure was to the inner circles of Cowles orthodoxy, the more loathe they were to apply explicit computational analysis to the logic of the constrained optimization of utility.

One might, like Simon, suggest that maximizing utility was empirically implausible; but figures such as Lewis and Kramer, who insisted it was mathematically computationally *impossible,* were rapidly consigned to utter obscurity.[33] This massive blind spot concerning computation was also distinctively constitutive of what "information processing" meant at Cowles in the later twentieth century.

Kenneth Arrow

In our Cowles roster, it was not always the person who innovated the most profound contributions who ended up wearing the laurels when it came to retrospective honors in the history of the economics of information. In any event, for the average orthodox economist (and the external spectator), it was Kenneth Arrow who eventually became the Cowles poster boy for an economics of information; and indeed, many of the themes covered here would be found in his work at one time or another: information as thing-like commodity, information as public good, knowledge flaws due to missing markets, cognition as intuitive statistics, tacit knowledge in the guise of learning-by-doing, decision theory as ersatz psychology, the Blackwell formalism, asymmetric information and moral hazard, bounded rationality, complexity theory, and even (a brief flirtation, quickly repudiated) cognition as computation. If one does not look too comprehensively at his *oeuvre*, one can find some modicum of support for just about any subsequent orthodox approach to the economics of information one might care to promote; and this may account for some of Arrow's popularity within the profession.

The irony of the profession's praise of this eclecticism is that, at one juncture or another, he has also repudiated each and every one of them.[34] The pattern seemed to be that whenever a particular research line concerning information threatened to invalidate some critical foundational aspect of the Walrasian program or other, Arrow would belatedly repudiate that research line and retreat. The one thing he never ever countenanced, however, was the primary notion that neoclassical models were an awkward, galumphing vehicle with which to express the primacy of the marketplace of ideas. This may explain some his recent crotchety statements, such as:

> The idea that people have difficulty computing the system has a long history; you can see it in Veblen, for example. But nothing followed from this insight. Herb Simon was a great apostle of this view. He's a great figure, and his work did lead to a research program, but in my view, it fizzled out.... As I think more

about complexity theory, I become more convinced that there is some sense we will never know how the economy operates.[35]

The mordant fact is, of course, in his old age, Arrow never sounded more like anyone else than Hayek.

Roy Radner was a minor Cowles figure who sought to ponder even more seriously the implications of cognitive science for the Walrasian program, exploring the observation that no agent should be presumed to engage in a trade that depends upon information not available to him at that juncture; he insisted that a Pareto optimum could only be defined relative to a given structure of information. Contradicting Arrow, he insisted that the separation between informational and computational considerations was entirely artificial, and wrote, "[t]he Arrow-Debreu world is strained to the limit by the problem of choice of information. It breaks down completely in the face of limits on the ability of agents to compute optimal strategies."[36] Radner's insights have been subsequently ignored for the most part, for reasons already broached.

Stanley Reiter

The final figure in our cast of mid-century Cowles Information Argonauts is Stanley Reiter (1925–2014). Now utterly ignored in the history of modern economics,[37] we think he deserves to be resurrected as a significant transitional figure in the Cowles landscape—from mechanism design to experimentalism in economics, from the Walrasian tradition to a particular conception of computationalism, and from the orthodox position that "all markets are alike" to something hinting at a diversity of market forms. In residence as a research associate at Cowles for only two years, 1948 to 1950, he ended up serving as the most devoted acolyte of Leo Hurwicz in the postwar profession, turning much of the

Purdue University economics department into a Cowles outpost in the period 1954 to 1967. Purdue was a powerhouse at this time, producing such students as Hugo Sonnenschein, John Ledyard, Nancy Schwartz, and Mort Kamien. More to the point, Purdue was a hotbed of Cowles-style Walrasianism when it was still of dubious general popularity; it included Hurwicz students James Quirk and Rubin Saposnik.[38] Because Hurwicz maintained close ties with the Purdue faculty, Ed Ames called him "an honorary member of the department."[39] But most significant for our narrative, Reiter in the period 1954–67 overlapped with Vernon Smith at Purdue, precisely when he was innovating his distinctive version of experimental economics. Reiter was the conduit through which Hurwicz's nascent mechanism design (and its Cowles flavor) came into conversation with the simplistic Marshallianism at the foundations of the experimentalism of the early Vernon Smith. In his memoirs, Smith later praised Reiter as "our leading economist" at Purdue;[40] but the intellectual connections can only be fleshed out by briefly considering Reiter's work.

Reiter started off at Cowles as a student of Koopmans's linear activity analysis, but that soon palled, and afterwards he lent his mathematical talents not only to formal statistics but also to problems of the dynamics of price formation. The latter issue brought him close to Hurwicz, even though his thesis advisor at Chicago had originally been Milton Friedman. He also demonstrated a fascination with algorithms for optimization under uncertainty—for instance, in the "job shop problem" in the early 1960s. By the 1970s, he announced himself as one of the first popularizers of the Hurwicz research program in mechanism design, retailing it as the "(New)^2 Welfare Economics," an awkward moniker that never really caught on. That paper stressed that informational considerations were central to the theory of economic mechanisms, positing that "An initial distribution of knowledge about the economy is assumed. Each agent knows something, but generally not everything. . . . No agent by himself knows enough to figure out the feasible allocations."

Even compared to Hurwicz, Reiter was much more insistent upon portraying the economy as "a kind of machine which accepts as inputs the basic data of the economy and produces as an output an allocation of commodities among the participants."[41] The computer as information processor overtly structured Reiter's account of mechanism design—to a much greater extent than many of other theorists of that era. This was exemplified by his work with Kenneth Mount on "The Informational Size of Message Spaces" and the restrictions that they imposed on Hurwicz-style mechanisms.[42]

Probably at Purdue, Reiter began to try to bring Cowles-style concerns to Vernon Smith's stress on market formats in his early experimental work. As acknowledged by Smith himself, he later attempted in his famous 1982 paper to undertake "the bridge-building . . . between experimental economics and the Reiter-Hurwicz—sometimes called the Northwestern—view of economic theory."[43] The computer, and the concern over information processing, was the unlikely common denominator of the two streams of Cowles-inspired Walrasian market socialism and Smith's Marshallian-style Hayekianism. Thus, we argue that Reiter turned out to be one of the key obligatory passage points between the earliest Walrasian mechanism design and the more modern subsequent market constructivism, often associated with experimental economics. But the full story must be postponed until a later chapter.

Reiter not only served as a vector of ideas but also played a major role in the reorganization of the American social sciences in the early 1980s. In a story first recounted by Kyu Sang Lee, there had been an attack on funding for the social sciences in the early Reagan administration, which provoked a response organized by the National Research Council and the National Science Foundation; they created a Committee on Basic Research in the

Behavioral and Social Sciences, launched in 1980. In an attempt to reassert the legitimacy of state funding of the social sciences, a number of reports were commissioned, the details of which will not occupy us here. However, the working group on Markets and Organizations was chaired by Stanley Reiter; and they produced a report which was extremely revealing about the contemporary state of the economics of information and its relationship to mechanism design when it appeared in 1989.[44] We shall make extensive use of this document in the rest of this volume.

Our impression is that Reiter moved further and further away from the Walrasian organon as he grew older, which may account for some of his neglect in orthodox retrospectives. For instance, in 2001 he made the extremely revealing admission that there were no real markets in the Walrasian model—something that would tend to disrupt the allegiance of a true follower of Hurwicz.[45] Even later, he conceded that there was still no plausible dynamics for Walrasian general equilibrium, and thus dabbled with simulation in agent-based models to explore other formats of convergence to equilibrium. In this late paper, he reveals he still seeks to make connections to the Vernon Smith school of experimentation:

> [T]he well-known informational efficiency of the competitive allocation mechanism is limited to the static model.... One line of response to this challenge has its roots in computer science. Developments in distributed computation have inspired research in which computational algorithms ... compute market equilibria or optima.... There is also a substantial literature reporting laboratory experiments in trading.... Experiments tend to show that the behavior of experimental subjects is roughly consistent with what economic theory assumes. It is not clear how much of the behavior observed in small scale experimental settings survives in a large economy.... Vernon

Smith and Charles Plott have pioneered this approach to studying trading.[46]

WHY COWLES?

Up until the move to Yale, the Cowles Commission sheltered the most amazing list of progenitors of the economics of information—at least in retrospect. Now that we have introduced the cast of characters at Cowles, we have to briefly return to the larger question: Why Cowles and not, say, MIT? Why was it they who set the templates for developments in the postwar economics of information? After all, MIT has been considered one of the three main centers of the diffusion of information theory in many sciences in the immediate postwar period.[47] In this regard, economics was an outlier.

There are a number of things to keep in mind about Cowles when observing various comrades foraging about for an economics of knowledge. First, because of their intimate connections with RAND, they were in much closer physical proximity to key natural scientists engaged in innovating new approaches to information than were any other schools of economics, neoclassical or otherwise. For instance, John von Neumann had made a number of overtures to Cowles economists in the late 1940s, which is why they were the first to entertain formal game theory. Kenneth Arrow in particular was a close colleague of David Blackwell, collaborating with him in the late 1940s;[48] Leonid Hurwicz and Stanley Reiter enjoyed close collaborations with various computer scientists. David Blackwell turns out to be an extremely important protagonist in this story: he was a mathematician who started out as an advocate of Bayesian statistical inference, but while working for the military at RAND, he came up with a novel formalization of information as measures over partitions of discernable states of the world.[49]

Second, as we have repeatedly stressed, many of the Cowlesmen explicitly admitted that their motivation in the 1940s–1960s in discussing information was to refute Hayek, and thus to show that information economics need not have neoliberal implications. MIT economics was so isolated in the 1950s that they didn't even realize Hayek posed a threat of some sort to their self-confident science. Whatever happened to be the postwar lay of the landscape, in the longer view it seems apparent in retrospect that the hunter got captured by the game—that is, the seductive frame tale of the omniscient neoliberal marketplace of ideas came to dominate much of their own work in mechanism design, asymmetric information, "failures" of expected utility theory, "incomplete markets," and a host of other innovations. Although Cowles as an institution decamped from Chicago in 1955 for Yale, the program it pioneered was continued at RAND, Stanford, Israel, Louvain, and wherever else operations researchers gathered together under military auspices. Perhaps more significantly, the Cowles approach later became ensconced in the postwar "new model" business school, partly in reaction to the rejected Harvard model.[50]

Third, in the modern orthodoxy, the primary visible heritage of Cowles for the median economist circa 1990 came with their latching on to the "state space" formalism as purportedly plug-compatible with their Walrasian general equilibrium orientation, a formalism first pioneered at RAND by David Blackwell. Since MIT was in principle agnostic concerning the cogency of full Walrasian general equilibrium, it tended to miss this particular ship's having sailed. Blackwell's centrality to a particular strain of microeconomics was summarized in the following obituary:

> Blackwell's approachability theorem is at the heart of Aumann and Maschler's result about repeated games with incomplete

information. ... Blackwell's theory of comparison of experiments has been influential in the game-theoretic study of value of information. ... Another seminal contribution of Blackwell, together with Lester Dubins, is the theorem about merging of opinions, which is the major tool in the theory of Bayesian learning in repeated games.[51]

In effect, information processing under the sign of Blackwell became confused with an image of the neoclassical agent as a little econometrician; one can observe this very starkly in some subfields such as rational expectations economics or in Bayes-Nash game theory. This is still currently treated in certain retrograde quarters as *the* "standard model" of information in economics.[52] We only want to insist that this model cannot be the one and only superior conception that has demonstrably stood the test of time—if only because induction should never be confused with the sum total of cognition.

[8]

THREE DIFFERENT MODALITIES OF INFORMATION IN NEOCLASSICAL THEORY

Although most of the initial action occurred at Cowles, it is imperative to ask where the formal conceptions of information appropriated by economists came from. Since information is both Promethean and polyvalent, there was no "right" place to look, no uniquely "correct" way to account for information within the existing neoclassical orthodoxy. Perhaps more to the point, there was no one right way to respond to Hayek. History is not usually constrained to run down a single track, which is why the history of the economics of information eventually turned out to be quite peculiar.

In this chapter we will briefly describe in an abstract manner the different modalities of information economics explored at Cowles, as summarized in the "main approaches" shown in table 8.1.

This does not exhaust every possible modality of modeling information in economics, but we will assert that it does manage to span the main lines of development in the latter half of the twentieth century. Moreover, we have another motive in proposing this taxonomy: when combined with our previous continuum of the different ways that Hayek explored in treating agent knowledge,

Table 8.1 THREE FORMAL APPROACHES TO INFORMATION

Information is:		
a thing (Shannon)	an inductive index (Blackwell)	symbolic computation (Turing)
Cognition is:		
irrelevant	intuitive statistics & epistemic formal logic	symbol manipulation
Learning is:		
purchase of a commodity	statistical inference	algorithm augmentation
Communication is:		
same as exchange	"signaling"	information transmission
Information generated in:		
The Past	The Future	The Now

we will append the three formal modeling options, thus to compose the two integrated axes to define the state space of the trajectory of orthodox information economics in the latter twentieth century. That will constitute the plot line for much of the rest of the book.

In the following sections, we shall peruse each column of table 8.1 in turn: information as thing, information as inductive index, and information as formal computation. Since the inductive index approach is so central to the story proffered in the remainder of this volume, it will take up a disproportionate space in this chapter.

INFORMATION AS A THING/COMMODITY

Treating information as some fluid but generic object was ubiquitous in postwar neoclassical economics—so much so that all manner of commentators would acknowledge the irony of the situation:

> Information is some sort of undifferentiated fluid that will course through the computers and telecommunication devices of the coming age much as oil now flows through a network of pipes; the measure of knowledge in the world will be the amount of info fluid we have managed to store up.[1]

If one dominant heuristic of postwar economists was "Do as little as possible to revise or alter the neoclassical theory handed down from our forebears" when discussing the operation of the marketplace of ideas, then one can readily appreciate why this option would have initially appeared so attractive. If information really is a thing-like object in nature, then it could just be subtended to the commodity space as one more good, and then apparently "nothing" need be changed about the standard maximization model. If information can be passed off as the stuff of knowledge, then all cognition can be banished; the received textbook model would be safe as houses (before the 2008 crisis!). Moreover, such a thing-like information concept would absolve the theorists of having to confront whatever model of mind was supposedly inherent in the utility function: and a good thing, too, since to render information thing-like simply presumes that the value *and relevance* of some particular bit of information could be automatically and unambiguously recognized by the agent in question, without any pesky need for interpretation or even a modicum of preparation or alertness.

Playing fast and loose with commodity space was a popular pastime in postwar neoclassical economics—think of the way Gerard Debreu deformed it to model uncertainty, or Kelvin Lancaster contorted it to capture "qualities"—and so editing "information" in this manner seemed a snap.

Information as a thing turned out to be far more slippery than its advocates expected, however. Although various versions were proposed beginning in the 1950s, the preferred options tended to gel in the 1960s, with Kenneth Arrow (1962) portraying scientific knowledge as a "public good"; Gary Becker (1964) lumping together knowledge, information, ideas, skills, and health of individuals, all under the rubric of "human capital"; and Fritz Machlup (1962) busily taxonomizing the information commodity into different types of "goods"—investment, intermediate, and final consumption. Later still, information deliquesced from solid to liquid, "spilling over" into all sorts of positive externalities.

The problem immediately arose as to how to "measure" or "quantify" this kind of information, and that is where Claude Shannon's "information theory" served a critical function. Shannon had developed an argument that suggested information could be treated just like entropy in physics, comparing it to an enumeration of the ways a stochastic microdynamics of symbols could make up a measurable macrostate of messages. A concept originally fashioned to discuss mechanical obstacles to communication channels may turn out to be utter nonsense when used to discuss the semantics of communication in trade, as many soon came to suspect. But that did not exhaust its significance for economics. The Shannon mania of the first two postwar decades had the unintended consequence of bolstering the general impression that scientists could and should treat information as a quantifiable thing, and even as a *commodity*. In practice, it became quite common to conflate the embodiments and encapsulations of knowledge in objects and artifacts as mere epiphenomenal manifestations of a generic "thing" called information. It was a reification based largely upon a misapprehension— but that didn't mean it still wouldn't have untold consequences down the line. One suppressed implication was the bogey of self-reflexivity: "I do not suppose that the information content of this essay could ever be quantified."[2]

Nevertheless, once knowledge was broadly construed as a "good," then arguments could begin over just what special sort of good it just might be. A riot of metaphorical invention ensued in economics. Arrow, in his military research in the field of operations research, took to referring to algorithms as forms of "capital."[3] Once freed of the cold war, the ill-defined concept of capital was deployed to indiscriminately refer to any generic investment in people, as in "human capital": this constituted the program of Bank of Sweden Prize winner and MPS member Gary Becker. No better instance of the backward-looking, old-fashioned construal of the substance could be imagined. Just stick the protoplasmic substance into a conventional utility function, or perhaps a production function, then stir, bake, and pontificate. Yet, when forced to define it, Becker always put "information" up front: "Human capital refers to the knowledge, information, ideas, skills, and health of individuals. This is the 'age of human capital' in the sense that human capital is by far the most important form of capital in modern economies."[4]

Or, perhaps its special conditions of production dictated its distinctive status as a "public good." If you could get people to accept that knowledge was an eminent instance of such a "good," it helped if you then began to endow it with all sorts of peculiar qualities. Starting with Paul Samuelson (1954) and Kenneth Arrow (1962), knowledge was claimed to be a weird sort of thing whose use by one person did not restrict or prevent its use by another (in the jargon, "non-rivalrous"); but also something from which it was intrinsically difficult to prevent another person from enjoying the benefits once you bought it (in the jargon, "non-excludable"). This created all sorts of problems for mathematical modeling, but more to the point, it was used in the 1960s–1980s to justify state subsidy and provision of this marvelous commodity.

We even can glimpse at how the endowing knowledge with thing-like qualities inspired the tradition of so-termed asymmetric information, with the famous "Lemons" paper by George Akerlof (1970). That paper has an MIT-style toy model that was used to argue a different rationale for government intervention—namely, that asymmetric information (as in the used car market) would cause only faulty cars to be offered on the market, because good cars were constrained to be sold at the same (Blue Book) price. If only the embodied car knowledge had been separately priced! The "bottom line" in that exercise that no cars would be sold at all had (of course) no relationship to the real-world used car market; but that was not the point of the exercise. As Akerlof argued in his Nobel lecture: [T]he study of asymmetric information was the very first step towards a realization of a dream. That dream was the development of a behavioral macroeconomics. . . . The modeling of asymmetric information was to price theory what the modeling of putty-clay, vintage capital and learning by doing had been to growth theory.[5]

It was often the case that when economists evoked information in the immediate postwar period, they relegated it to the category of "public good" as a prelude to simply presuming it a thing-like commodity. The neoliberals had, of course, attacked the concept of knowledge as public good, especially those members of the "public choice" persuasion, more often than not through the instrumentality of the Coase theorem—the deconstruction of thing-like information in economics now is counted as one of the fabled contributions of MPS member Ronald Coase. But upon the heels of the neoliberal turn, a curious scholastic argument was subsequently made that the previous characterization had been mistaken, and that knowledge was only "partially excludable," and was distinctively different from "human capital," rendering it an even more special category beyond "public goods."[6] This ontological slipperiness of what, after all, is supposed to be a physical "given" to the model, is the first symptom of an outbreak of radical indeterminacy in this particular approach to an economics of knowledge.

No advocate of these models ever proceeded to resolve the "information" involved into actual measurable "bits"; neither did anyone go about modeling a "channel" with the normal Shannon characteristics of a fixed capacity, or a noise source. Moreover, no real-life market sold anything like commodity units of "information"; every real-life application involved sale of some other derivative object (a book, a lecture, an experience) or a set of legal property rights. As fictional stylists, economists betrayed a weakness for synecdoche, misrepresenting the part for the whole. They tended to conflate intellectual property with information, even though doing so exhibited a severe misunderstanding of the nature of patents.[7] Any such objections were treated as mere quibbles; "information" was pronounced the lifeblood of the New Economy, and nothing would withstand the drumbeat of the reification of information into a commodity.

INFORMATION AS AN INDUCTIVE INDEX AND/OR THE STOCHASTIC OBJECT OF AN EPISTEMIC LOGIC

With the development of mathematical statistics, there had been efforts early in the twentieth century to link intuitions of a "good sample" to the amount of "information" it contained, particularly in the tradition of R. A. Fisher. However, none of these proposals amounted to much outside the ambit of a small coterie of statisticians. However, in the postwar period, an interesting phenomenon happened whereby the statistical tools of inductive inference (having just spread throughout the social sciences) began to get conflated with models of mind.[8] Since the story of psychology in the early twentieth century consisted of a series of frontal assaults on the conscious mind as executive in charge of rationality, a revanchist movement resorted to the theory of probability to stem the tide.[9] The situation changed rather radically when mathematical statisticians were brought together with operations researchers and game theorists at the RAND Corporation in the early 1950s. There, especially in the work of David Blackwell, a practice took hold of equating "information" with measures defined over partitions imposed upon an exhaustive total enumeration of states of the world, both actual and virtual.[10] Crudely, how much a procedure (it was harder to phrase this in terms of real people) "knew" about a world was a function of how finely or coarsely it could divide up the possibilities, distinguish the class of outcome, and thus assign probabilities to eventual outcomes, as well as the sensitivity with which its detectors could discriminate which of the possibilities had actually obtained. The necessity for game theory to divide and discriminate strategies according to states of the world was an immediate inspiration, but quickly the formalism was developed in two, relatively separate directions: one was as the framework for modern definitions of one version of inductive inference, and the other was as the scaffolding used to assign semantic relations to a modal logic.[11] In an alliance with artificial intelligence, it became the basis for formal models of an important class of machine logic. In retrospect, this turned out to be the most machine-like construction of the nature of knowledge within orthodox economics.

When students are told that an economics of information has "solved" the problem of asymmetric information, more often than not the speaker is making reference to a complex of modeling practices, which we will call, in honor of its progenitor, the "Blackwell program." This approach also began outside of economics—in the first instance, among operations researchers and theoretical statisticians (the most eminent being David Blackwell)—but was rapidly taken up by the economics profession after 1965. The model consists of four basic components, with others added to adapt the model to the specific circumstances (but bypassed here).

Component 1: Reality is a Pre-Existent State Space of Possible Worlds

"Knowledge" in this approach is said to consist of partitions over a totally exhaustive state space, which distinguishes possible worlds by the inclusion or exclusion of a preset menu of variables. Greater knowledge is said to be represented by finer and finer partitions over the invariant state space, allowing greater precision concerning location in the space. Notice, the ontology of this world is given and fixed prior to the analysis, and cannot be altered by any activities of the knower. Time, by construction, merely is one of the state variables. Thus, this rather resembles the "block universe" of relativistic physics, where past, present, and future all coexist simultaneously in this state space. This is represented in figure 8.1. One immediate implication is that "signals" might emanate from the future, as well

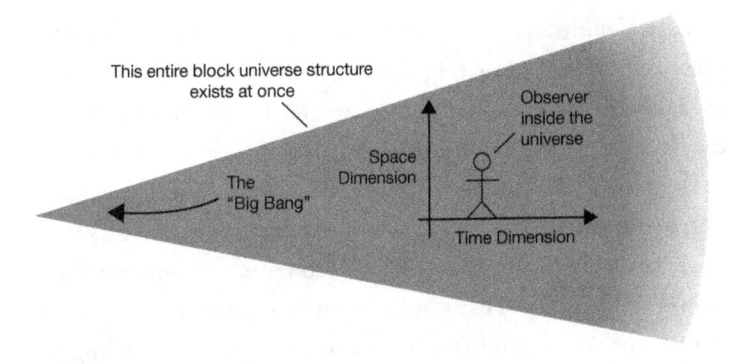

Figure 8.1. The Block Universe of Relativistic Physics.

as the past. Readers should note that nothing really new or unprecedented can ever appear in this space.

Years of experience have taught that the state space formalism models a perverse sort of "pre-cognition" built into agent consciousness: in effect, agents are presumed *ex ante* to know of the existence and relevance of as-yet-unlearned information; thus, there is no room for Hayekian "unknown unknowns" here. It has been formally proven in the interim that agents in state space models cannot be said to be unaware of anything permitted in the block universe; furthermore, the agent can never "know" that he is in fact unaware of any event or consequence. This seems to take much of the air out of the Aumann-defined pretense of common knowledge, described later.

Component 2: All Information Is Inductive

Blackwell proposed that "experiments" consisted of information extracted from the state space by inductive inference of probabilities attached to partitions of the state space. "Information" consists of signals emitted from experimental interventions, which serve to alter epistemic probabilities defined over the space. In theoretical statistics, Blackwell (1953) compared this to a "game against Nature," which reveals the formal inspiration from postwar work done on game theory, particularly in the formal analogies with "strategies" and "payoffs."[12] Blackwell himself was partial to the statistical school of Bayesian inductive inference, with the scientific researcher beginning with "prior probabilities" inherited from the past.

Although early game theory had originally come equipped with no psychological or epistemic capacities whatsoever, the mathematical similarities between the Blackwell formalism and game theory induced game theorists to explore this approach to knowledge and information in their quest to generalize its ambitions. The other major application of the theory came under the rubric of "machine logic" or "epistemic logic," primarily for use in computer programming.[13] When developed in computing, the Blackwell structure was used as a convenient model of induction, and not as some general approach to human, or even artificial intelligence.

The same could not be said for the economists. It was at this stage that neoclassical economists began to conflate any problem of "information" with some version of generic "choice under uncertainty."[14] As Thomas Schelling once said in 1962, "There is a tendency in our planning [models] to confuse the unfamiliar with the improbable." Any epistemic problem of any stripe, in pursuit of analytical tractability, was to be reduced to a set of probabilities induced over utility payoffs. As economists sought to incorporate this structure within their prior models of given preferences and von Neumann–Morgenstern expected utility, they discovered that the absence of cognitive content wreaked havoc with the result. Little toy models would posit as an assumption that some agents purportedly "knew" something that others did not, denoting this as an example of "asymmetric information," but the mathematics proved so arbitrary as to verge on emptiness.[15]

Once one opens up this scheme to doubt, it can rapidly rot the entire model enterprise. For example, should the relevant state space formalism include not only states of nature but also the states of mind of rival players? (Recall from Component 1 that this would imply subjective states of mind should be included in the changeless block universe.) What would validate the truth of things believed to be "known" by respective payers? Nothing in existing utility theory permitted the formalization of the infinite regress of "I think that you think that I think that you think that . . ." (unless all players were effectively identical, so knowledge is still "perfect," as in early Nash non-cooperative games). Other subtleties popped up in the machine learning literature: "Although you may have false beliefs, you cannot know something that is false."[16] A parallel false universe was banished by construction.

Component 3: The Harsanyi-Nash Program

A number of alternate responses to the conceptual problems presented by the Blackwell setup could have been explored; some economists admitted this in the early stages of the inductive approach to information: "What equilibrium is in a particular market depends on what individuals in that market know. That the converse is true—that is, that what people know (or believe) is a function of the equilibria of the markets in which they participate—is an observation which surely must precede Marx."[17] However, very few economists had the stomach to explore the ways knowledge and markets interactively shape one another in an alternative approach to microeconomics.[18] Instead, by the 1970s, the preferred exit out of this conundrum for game theorists involved the uncritical preservation of the standard formalism of von Neumann–Morgenstern expected utility, combined with uncritical adherence to the Nash solution concept in game theory. The way this is often put is that all imperfections of information about the world in the guise of uncertainty in games must be reduced to uncertainty over parametric payoff functions. The person who supplied the preferred escape route was John Harsanyi.

Harsanyi developed his approach to information in conjunction with other game theorists, such as Robert Aumann, while employed by military agencies to apply game theory to problems of nuclear war and disarmament. As he saw it, the existence of "incomplete information" would lead game theorists to have to incorporate an infinite hierarchy of beliefs into their models of agency. These would consist of a pyramid of beliefs for player i over parameter vector X in the state space consisting of:

First-order beliefs: Player i's probability distribution over vector X
Second-order beliefs: Player j's probability distribution over player i's first order beliefs
Third-order beliefs: Player i's probability distribution over Player j's second order beliefs
... And so on, ad infinitum.

An attempt to reduce this disturbing situation to a single picture is portrayed in figure 8.2.

Clearly this infinity appears daunting (not to mention the cognitive demands it imposes upon strategic choice). Harsanyi proposed to recast the model to "define away" these infinities (perhaps similar to the way renormalization dispensed with infinities in the case of quantum electrodynamics) by reducing all problems in epistemology to problems in the definition of the agent.[19] He argued for

Investors base their decisions on expectations about future economic policy.

Central banks set the interest rate based on expectations about private sector developments.

Figure 8.2. The Harsanyi Setup.

this in his (1967) paper by suggesting that *anything a player knows "privately" that could affect the payoffs in the game about other players* could be summarized in a single vector, called indifferently the player's *information vector* or *type*. This "simplification" in effect collapses everything that a player knows privately at the beginning of a game that could potentially affect his beliefs about payoffs, plus the "types" of all his opponents into a single roster of typology of players.[20] "Information" disappears as a discrete analytical entity, only to be replaced by an artificial zoo of player "types," or as textbooks often put it, any game of incomplete or missing information becomes a game of imperfect information—the only uncertainty is over which automatons "Nature" has bequeathed you as opponent. The artificiality of this model strategy has been acknowledged repeatedly by its second generation of advocates: "we use type structures solely as a modeling device. Types are not real world objects."[21]

Harsanyi's model device renders formal the vernacular maxim: It's not what you know, it's *who* you know. To preserve the Nash solution concept, it is the type of actual players that is the primary unknown in the analysis; the roster of types, along with the structure of the game itself and the rationality of the players, is given a priori and presumed known to all players. Since the theorist supposedly places himself on the same epistemic plane as the players, the only way to learn anything further about the game is through Bayesian inductive inference. This is the current meaning of information buried at the heart of Bayes-Nash game theory. The standard game setup ends up inverted; realized payoffs tell you who your opponents really are. Of course, once mixed strategies are allowed over types, then all meaning of player identity dissolves into thin air. How you are supposed to know who you yourself really are under such circumstances is a mystery.

Component 4: Common Knowledge

It was left to Harsanyi and Robert Aumann to draw out the final implications for knowledge in this marriage of orthodox game theory and Blackwell formalism. The irony is that a modeling approach that sought to place inductive approaches to information on a sound theoretical footing ended up banishing ignorance altogether, at least for the Bayes-Nash agent. The involution began with knowledge of the block universe, presumed recognized as true by all participants, incapable of expressing falsehood. It intensified with the Harsanyi procedure of presuming that players also come equipped with full identical knowledge of the game they are playing, which includes the roster of all possible player types, and shared recourse to the practice of Bayesian inference.

Given the extensive shared knowledge presumed on the part of all players, Harsanyi realized that it would be arbitrary to allow different "players" to start the game equipped with different Bayesian prior probabilities, since that would constitute the only thing that would be "unknown" to opponents in any deep sense. The entire thrust of the Harsanyi program was to banish from the standard model any parameters that are not already "common knowledge" among all players; so, therefore, he propounded the doctrine that player types also came equipped with identical Bayesian priors.

At this point, the Bayes-Nash approach to information disappeared up its own navel as a consequence of its own insistence upon rigorous consistency with its own postulates. Robert Aumann (1976) originally used his definition of common knowledge to prove a notorious result that says that, in a certain sense, agents cannot "agree to disagree" about their beliefs, formalized as probability distributions, if they start with common prior beliefs. Since agents are often conventionally portrayed as holding different opinions plus some beliefs they do not hold in common, one might attribute such differences to the agents' having different private information. Aumann's incongruous result is that even if agents condition their beliefs on private information in a Blackwell setup, mere common knowledge on their part of their conditioned beliefs plus a common prior probability distribution implies that their beliefs cannot be different, after all.

This seeming travesty of the economics of information has given rise to a cottage industry exploring the meaning of such locutions. One important student of Harsanyi has admitted that "[t]here is something fundamentally counterintuitive about the art of modeling [information] with Bayesian games,"[22] and reported that Harsanyi himself sometimes seemed uncomfortable with the implications of common knowledge in his approach. Other economists, less concerned with epistemic niceties, regarded the Harsanyi setup as license to apply the model to all manner of "information asymmetries" in the phenomenal world. Their version of "common knowledge" was interpreted to mean everyone who was "rational" had to agree with their model.

INFORMATION AS COMPUTATION

This version of knowledge owes the greatest debt to the postwar development of the computer and the theory of computation, but curiously enough, has proved over time to be the least palatable for many neoclassical economists. It predominantly travels under the banner of "computationalism," which tends to identify mental states with the computational states found in (either abstract or tangible) computers. Computationalism comprises many competing visions, ranging from formal symbol manipulation, to "connectionism," to "machine cognition"; but economists have rarely been very sensitive to these controversies within artificial intelligence and cognitive science. For instance, economists rarely realize that the "connectionists" and the proponents of genetic algorithms often praise Hayek as an early progenitor, whereas the first generation of artificial intelligence theorists and complexity mavens instead identified Herbert Simon as their inspiration.

To simplify our exposition, here the processing of "information" is equated with symbol manipulation by automata of various computational capacities, with the Turing Machine occupying the highest rung on the computational hierarchy. In a ranking of the power of various abstract "machines," the class of Turing Machines are generally conceded to be the most powerful.[23] The importance of the computational hierarchy is that it facilitates the proof of impossibility theorems concerning what can and cannot be computed upon machines falling within a particular computational class. Computational approaches have had the prophylactic virtue of ruling out all sorts of physically and mathematically impossible procedures from falling within the purview of an algorithmic conception of rationality. Treatment of infinities assumes much heightened significance; implementable algorithms are more highly regarded than in-principle proofs.

The computationalist turn has assumed two different formats in the history of orthodox economic theory: the first attempts to subject the standard rational choice model to be subsumed under a computationalist model of mind, while the second tends to fall under the rubric of "market design." The former cadre are a rather diverse lot, ranging from those (such as Alain Lewis) seeking to model individual rational choice as an explicit computational proposition, to what has been called "agent-based computational models,"[24] but might be better thought of as simulations of agent swarms, after the model of cellular automata.

Actual experience with computers has provided all manner of heuristic suggestions as to how to meld cognitive science with neoclassical economics, perhaps taken to an extreme at certain locations. Indeed, as one Clark Medal recipient has admitted, "if you try and do psychology at MIT, you study computers, not humans."[25] The latter market design wing combines certain sectors of experimental economics with what might be best described as "engineers of automated markets," where both claim to have superior insight into the informational properties of markets with large numbers of participants. This latter group has ambitions to be engineers of the human soul, arguing that their purpose-built machines can force people to tell the truth even when their every intention is to be mendacious, or provide them with information that they would otherwise find inaccessible through any conventional recourse to research channels. The market design tendency is covered in later chapters.

The role of Cowles in the exploration of the computational approach to an economics of information has been rather limited, compared to the other approaches. After Alain Lewis attempted to apply computational audits to the neoclassical agent, and this line of inquiry was rejected by Kenneth Arrow, the computational approach to mind has more or less been avoided by Cowles-related researchers. Conversely, the self-conscious computational approach to market automation also had few takers; Stanley Reiter came closest to that approach, but never really converted wholeheartedly. Herbert Simon also noticeably could not accept a formal computational approach to markets. Reiter, however, did provide a bridge to the experimentalists, and that is where we look to uncover the economic actualization of computational themes.

Early on, the computational metaphor of mind proved a mixed blessing for economists. If one were to seriously entertain the notion of a marketplace of ideas, the problem became where in the economy one would situate the computer. Was each agent a Turing Machine, or perhaps an automaton of less exalted capacity? The von Neumann architecture built into every laptop did seem a bit removed from human cognition, and then there were the interminable disputes of the 1960s–1990s over what it was that humans could do that computers could not. Most would admit computers could contain and manipulate information, but could a computer be seriously thought to be knowledgeable?

The development of the Internet seemed to present templates for the formalization of the communication of information. Or possibly edging closer to Hayek's vision, perhaps the marketplace itself should be treated as one vast Turing Machine, with agents simply plug-compatible peripherals of rather diminished capacities? This problem was compounded by the patrimony of the original neoclassical model, located as it was in non-computable N body mechanics.[26] The history of this research program reveals that certain aspects of the neoclassical model were shown to be Turing non-computable.[27] The temptation was then to shift the location of the computer to another ontological level in order to evade the unsavory implications.

Nevertheless, even in the face of that incompatibility, there is a sense in which the intellectual sway of the computational approach to information and its processing was unrelenting, and therefore inescapable for the neoclassical orthodoxy. Strangely enough, one place this is readily observable is in experimental economics. The practical requirement to program little scaled-down toy economies in computerized laboratories as environments for the subjects to inhabit rapidly led to the extension of computational considerations in all manner of questions broached by those experiments. We can illustrate this by looking at one exemplary case, that of MPS member and Bank of Sweden Prize winner Vernon Smith.

Vernon Smith and his Smart Markets

Vernon Smith, as he himself testified,[28] produced his first experiments in reaction to Edward Chamberlin's Harvard classroom exercises that purported to show that free competition would not produce efficient exchange outcomes. Smith bypassed psychological experimentation (which had tended to refute neoclassical models of agency) in favor of market simulations predicated upon rules derived from stock exchange manuals. His first experimental article (1962, in Smith 1991) demonstrated that in the presence of what became known as "double auction" rules, prices and quantities rapidly converged to supply/demand equilibria, even in the absence of any knowledge of the theory or data on the part of his subjects (usually university undergraduates). His early success led Smith to explore a range of other market formats instantiated as alternative rule structures in his laboratories; in conjunction with his colleagues, he found that no other type of market regularly produced what they considered to be superior outcomes—namely, rapid convergence to predicted price and quantity equilibria and near-full realization of the predefined consumer and producer surplus in his experimental setups. (In contrast with Pareto optimality, both objectives were computable by construction.)

One lesson he drew from this line of research was a thesis he dubbed the "Hayek Hypothesis" (1982, in Smith 1991): "Strict privacy together with the trading rules of a market institution are sufficient to produce competitive market outcomes at or near 100% efficiency"—that is, independent of the cognitive abilities or status of the agents involved. Smith believed that the neoliberal power of The Market had been proven in the laboratory, infusing it with the aura of real science.

Another lesson grew out of the move to automate his experimental laboratory through integration of computer technology in the mid-1970s. In his own words:

> Science is driven more fundamentally by machine builders, than either the theorists or experimentalists.... Computer/communication and imaging technologies are driving experimental economics in new directions. Both will marginalize extant research on individual decision. When Arlington Williams programmed the first electronic double auction (e-commerce in the lab) in 1976, it changed the way we thought about markets, much as the internet is changing the way people think about doing business. Circa 1976 we thought going electronic would merely facilitate experimental control, data collection and record keeping. What we discovered was altogether different: computerization vastly expanded the message space within which economic agents could communicate at vanishingly small transactions cost. This enabled Stephen Rassenti to invent the first computer-assisted combinatorial auction market.... Lab experiments became the means by which heretofore unimaginable market designs could be performance tested.[29]

So early experience with the computer automation of experiments, which meant in Smith's case experience with computer simulation of various market operations, prepared the way for Smith-style experimentalists to be among the first researchers with the requisite skills to program and implement never-before imagined variants of electronic markets. (Stan Reiter was one of the very few Walrasian theorists to entertain similar considerations.) The computerization of experimentation also had more than a little to do with Smith's predisposition to focus upon the double auction format to the exclusion of other forms; more than almost any other species of market, it was amenable to full reduction to an algorithm. These specializations in the coding of markets and favorable inclinations toward auctions, of course, will explain the preponderance of experimental economists to be found occupying the nascent field of "engineering economics." But more to the point, it also encouraged a more direct research initiative on the part of others even less committed to the neoclassical program than Vernon Smith himself to develop an analytical rationale concerning the relative independence of the algorithmic market from the neoclassical agent.

While there were a number of conceptual problems with Smith's enunciation of his "Hayek hypothesis," the most nagging was the query: If the success of the double-auction format was truly not dependent upon the cognitive states of the experimental subjects, then what did account for it? This question was posed and answered brilliantly by two Carnegie researchers, Dan Gode and Shyam Sunder. In a now-classic paper (1993), they compared the experimental outcome of a double-auction setup using human subjects with a fully automated setup that replaced the subjects with random-number generators, which they dubbed "zero-intelligence agents" (henceforth ZI). While there is still substantial dispute over the interpretation of their results,[30] it appeared that brainless programs produced nearly indistinguishable results with regard to convergence and efficiency compared to Smith's human subjects.

The computational insight of the "zero-intelligence" exercise was that human cognitive capacities could be zeroed out under controlled circumstances (thanks to the prior automation of experimental economics) in order to explore the causal capacities of markets conceived as freestanding automata. The predictable regularities of the double auction in experimental settings (and perhaps "in the wild," as experimentalists were wont to say) should be attributed to their algorithmic structures, and not to any psychological predispositions of their participants; as Sunder himself put it, "a science of markets need not be built from the science of individual behavior. . . . Market institutions may be society's way of dealing with human cognitive limitations. . . . Efficiency of markets is primarily a function of their rules."[31] In Gode and Sunder's subsequent work, they deconstructed the auction down to its algorithmic parts and subjected each component to the ZI trader test in order to further explore the sources of the efficacy of the double auction.

Here, then, was a formal harbinger of a neoclassical economics without the trappings of "rationality."

CODA

Thus, we have provided the outlines of the three variant forms of "information" explored in orthodox economics in the second half of the twentieth century. In one desiccated conception of history, this roster would be sufficient to encompass the obsessions of economists: an endless sequence of Model A—Model B—Model C—. . . . But the history of economics is much larger, and far more interesting than that. Just like the bumbling spies in our first chapter, as our intrepid researchers tugged on the string of "information," they inadvertently prompted their own world to unravel.

[9]

GOING THE MARKET ONE BETTER

In chapter 7, we saw how Friedrich Hayek's argument against socialism served as the initial provocation for economists to come to grips with "information." Economists at Cowles interpreted Hayek as arguing the relative merit of "free markets" over socialism on informational grounds, and they found this argument wanting. In a move that would have vast and enduring ramifications for the future of the economics profession, Hurwicz and his colleagues at Cowles responded to Hayek's provocations by reconceiving their task as external evaluation of the informational properties of economic systems, claiming soon thereafter that these methods could also inform choice among a plethora of "institutions."[1] The Cowlesmen eventually rebranded themselves as experts in "organization," a term that assumed brash capacious dimensions so as to cover such varied phenomena as the internal structuring of large companies, the design of cost-plus contracts for the mobilization of industry during wartime, the evaluation of Soviet central planning algorithms, and the crafting of regulation. Indeed, the historian Hunter Heyck has described how a fascination with "organization" became conflated with themes of algorithmic reason and analysis of information across the immediate postwar social sciences.[2] With increasing frequency, these new organization theorists (disproportionately concentrated at Purdue, Caltech, Arizona, and Northwestern) began to contemplate designing *new* institutions, ranging from novel legal regimes to "solutions" for public goods provision to the reorganization of entire economies.[3] And in what turned out to be the most significant development for the future of the economic profession, they would also claim an ability to reconstruct individual precursor markets themselves.

The Cowles pretensions to evaluate organizations in general thus paved the way to what eventually became known as "market design." There exist a handful of attempts to situate market design within (or in very close proximity to) neoclassical economic theory, usually by identifying a few select mathematical "tools" commonly held in the economist's kit, such as the envelope theorem.[4] But these inevitably obscure the big picture; whereas the neoclassical approach stressed the existence of a generic and omnipresent market, uniform in its qualities at all times and places, market design was predicated on a repudiation of this central imperative. Questions that would have verged on the incomprehensible if posed from within the previous framework (e.g., What laws or constraints govern the (design of) internal processes of organizations, including markets? What laws or constraints govern the endogenous adaptation of organizations to changing environments?) now guided the Cowlesmen, their students, and those in their orbit selecting topics for further research.[5] In short order, economists began turning out a variety of hitherto undreamed-of market devices, tailored for a variety of purposes—for example, space shuttle payload launches, airline scheduling, and medical residency assignments.

What could possibly be responsible for so momentous a change? We are fortunate in this case to have the real-time reflections of several of our major protagonists, in the form of a 1989 report of the Markets and Organizations working group of the Committee on Basic Research in the Behavioral and Social Sciences, whose members included Stanley Reiter, Kenneth Arrow, Roy Radner, and Leonid Hurwicz, among others. Therein one finds a key acknowledgment:

> We are now in the middle of a profound change in the technology of information processing. In the light of the importance of information processing in organizations, this change constitutes a profound change in the technology of organization itself.
>
> As research moved from auctions in which no information processing is involved to auctions that involve a heavy information component, paradoxes [such as the winner's curse] begin to emerge. . . . If the phenomenon persists under close examination, theorists will be forced to search for a modification of basic principles that are now widely applied. In addition, a search will be initiated for institutions that prevent what can be perceived as a problem.[6]

In short, *information* changed everything. It inaugurated a revision in what economists saw fit to analyze and what was falling within their remit. The multiple chickens in chapter 8 had come home to roost. This was no minor change to Walrasianism; attending to information would eventually necessitate further adjustments, not only to "basic principles" but also to economists' practical relationship with the economy itself—to the very purpose of economics as a vocation. No longer would the economist heedlessly berate the agent for failing to act rationally (oil companies in the case of the winner's curse, for following naïve bidding strategies[7]); now, economists in good standing would fix markets to make things happen.[8] No economist prior to 1980 had ever made such a claim. Hence, the final consequence of the introduction of information was to change what an economist *does*.

But who wanted this new species of economic expertise—and to what ends? Although the Markets and Organizations report provided an important endorsement of market design, it would be prudent to point out that such work was at that time still thin on the ground, and largely the province of a handful of tinkerers concentrated well beyond the groves of the Ivies, with many located outside of economics departments proper. So let us pan out a bit, beyond the bounds of the profession, to consider the circumstances surrounding the drafting of the Markets and Organizations report.

The time was the 1980s; in an age of retrenchment, the Reagan administration had proposed drastic budget cuts for social science funding. Understandably, this alarmed a broad range of social scientists and provided an impetus for them to join in an unusual effort to justify the practical usefulness of their disciplines and highlight the most promising areas for development (and funding).[9] With research financing now imperiled, the spadework undertaken at Cowles (and later, Purdue) began to bear fruit as the profession rallied around the study and design of novel "organizations." Slowly at first, but then with astonishing rapidity, the aspirations of economists and policymakers converged on the task of thoroughly redesigning the organizations of the economic lifeworld from bottom to top. Market designers offered to lend their expertise (for a price), and with increasing frequency policymakers took them up on their offer. This convergence was no accident. Policymaker and economist alike came to appreciate how real-world markets came to increasingly resemble information processors, and they adjusted their aspirations in light of this;[10] both would come to attribute immense epistemic capacities to these markets. As a practical matter, this justified the piecemeal marketization of government functions and ultimately full privatization; as a theoretical matter, this served to degrade the cognitive capacities of the agent. In a way we will explicate in the remainder of the book, market design turned out to be a perfect "fit" for its time, because it constitutes the precepts of neoliberalism taken to their logical conclusion.[11]

Like detective stories, histories of economics tend to conform to certain conventions that make the reader comfortable. Almost all current intellectual history of economics portrays its protagonists as if they were all talking about the same "thing," as if in a dialogue across generations. That endows the plot line with solid chronology and infuses a confident sense of cumulative understanding. For historians covering the post–WWII era, the object of their rational reconstructions has been the *agent*. The transformation of agency takes center stage for Floris Heukelom (2014), for instance, who offers an account of economists rejecting general characterizations of agent behavior in favor of descriptions grounded in observation. Before, economists contented themselves by operating with mere assumptions about agency; now such characterizations must be submitted to an empirical audit. Nicola Giocoli (2003, 2009) also focuses his account on agency, only he posits that the past several decades have been a period of regression. Agency underwent reconceptualization from maximization to formal consistency; as a result, the profession relinquished concern with the process by which agents pursue their desires. Eventually, Bayesian rationality caught fire, allowing economists to engage in institutional design.[12]

Perhaps hazarding a bit more controversy, we believe we can detect in these works a normative lesson. For Giocoli, the imperative of revising our representations of agency to correspond to the insights of psychologists (among others), to develop more realistic accounts of behavior and learning, would serve to "reaffirm the central role of human beings in economic models."[13] For Heukelom, whose vantage point is ten years later than Giocoli's, the stance becomes something like this has already taken place. The improved understanding of agents' behaviors, coupled with a more thorough understanding of normative theories of rationality, has provided economists with the means to help agents act more rationally (think here of Sunstein and Thaler's *Nudge*), and the moral obligation to do so. In heralding the Grand Emancipation of the agent, these histories resemble nothing so much as introductory chapters from textbooks on behavioral economics.[14] Rendering our agents more fully human will liberate us from the machine.

Generally, the rational reconstruction approach to history is misleading; but when it comes to the treatment of information, it ends up lost in Bedlam. What do we offer in its stead? We must admit there has been no single unified story here related so far but, rather, the intersection of a number of very big intellectual crosscurrents: the rise of information in the twentieth-century natural sciences, the consolidation of neoliberal political economy, the Socialist Calculation Controversy, Hayek on epistemology, the response of market socialists at the Cowles Commission, the selection by economists of a range of incompatible model options to incorporate into their neoclassical tradition, and the rise of mechanism design. These things all had some bearing upon one another, but as yet we have not directly made the case that there is a coherent story to be told about these events.

In the 1970s, information broke out of the narrow arenas we have visited in chapters 1 through 8, and infiltrated every corner of economics: micro, macro, econometrics, experimental, behavioral—you name it. Making a checklist of the highlights would simply compound the sense of arbitrariness in the reader; it would resemble a wild goose chase. Rather than that, we do have a story to tell with a proper plot line, and we believe it is of utmost importance for the shape of economics in the twenty-first century. Plainly, and starkly, the spread of the "information" concept throughout economics after WWII thoroughly and irreversibly changed the way economists thought about markets, and that happened in parallel with changes in actual markets themselves. Given that The Market as icon and touchstone remained so central to the self-identity of the economics profession, ultimately this dynamic changed what it meant to be an economist: it eventually altered what economists *did*. Their remit, primarily, shifted from market description to market *design*. This had nothing to do with a conventional progress narrative, which would track various protagonists working their way ever closer to the truth concerning a fixed empirical target. Our alternative, we insist, is a tale full of sound and fury, signifying quite a lot.

Every Aristotelian arc requires a pivotal protagonist. We opt to begin our discussion by identifying a single exemplary figure through whom one can observe these big trends at work. But whom to select? Such a person would have to be a member of the professional orthodoxy in good standing (unlike, say, Friedrich Hayek), situated at the center of attempts to grapple with the disruptive forces of information and ideally celebrated as a pioneer for his or her attempts to do so. Since the effects of information were only dimly perceived, and only belatedly, one should not be surprised if affirmation of this person's exceptional standing came unusually late. Moreover, if we are right in insisting that information did not remain a fixed target but, rather, was a movable feast, our coming to such late appreciation of his contributions would mean we would have to successfully navigate some hairpin turns. Consequently, we might expect a lingering and barely suppressed confusion about the coherence of our figure's central intellectual project.

We open this portion of our case by focusing on someone who fits this description to a T, the protagonist whom we cited at the outset of this chapter: Leonid Hurwicz.[15]

HURWICZ AT THE BIRTH OF MECHANISM DESIGN

Not so long ago, Leonid Hurwicz would have, at best, been regarded as a supporting actor in the history of economics—a collaborator of Kenneth Arrow's on some important papers on mathematical programming and the stability of a general economic equilibrium.[16] He would not have rated as a star, but maybe a second banana. Yet how things have changed. By now, the profession has come to attribute to Hurwicz something approaching an indispensable leading role, for "laying the foundations of mechanism design," in the words of the Prize Committee of the Royal Swedish Academy of Sciences, who awarded him its 2007 prize.

The reappraisal came very late. Although the Bank of Sweden Committee (BOS) has a well-earned reputation for selecting its recipients at a relatively advanced age, at ninety Hurwicz was its oldest ever laureate (the average recipient age has been sixty-seven). That he shared his prize with researchers over thirty years his junior (the game theorists Eric Maskin and Roger Myerson, among the prize's *youngest* ever recipients[17]), necessitated a highly circumscribed and carefully worded acknowledgment of Hurwicz's contribution. In its "Scientific Background for Mechanism Design," the Bank of Sweden Prize Committee eliminated any substantive reference to his work published prior to the age of fifty-five, a body of work that included the 1960 paper in which he introduced the concept of mechanism design.[18] Instead, it proclaimed "Mechanism design theory became relevant for a wide variety of applications only after Hurwicz (1972)" (Royal Swedish Academy of Sciences 2007, p. 2). The committee made mention of only those works directly relevant to those of Maskin and Myerson—which is to say, the Bayes-Nash approach to studying economic institutions; it studiously avoided mentioning Hurwicz's early *rejection* of the usefulness of game theory for this line of inquiry.

The Bank of Sweden Prize citation serves as a pretty fair approximation of the way the orthodox profession views the significance of Hurwicz's contributions. Jerry Green put it like this: "Hurwicz gave us the definitions, and we went to work."[19] Leonid Hurwicz: giver of definitions. The year prior to their joint award, Myerson provided a more expansive discussion:

> In an influential paper, Hayek (1945) argued that a key to ... economic theory should be the recognition that economic institutions of all kinds must serve an essential function of communicating widely dispersed information about the desires of different individuals in society. From this perspective, different economic institutions should be compared as mechanisms for communication.... The pivotal moment occurred when Hurwicz (1972) introduced the concept of incentive compatibility. In doing so, he took a long step beyond Hayek in advancing our ability to analyze the fundamental problem of institutions.[20]

Myerson's account—prepared for the Econometric Society as its "Hurwicz Lecture," explicitly to honor Hurwicz—omitted mention of anything Hurwicz wrote before the age of fifty-five. Researchers, beginning with Barone, Lange, Mises, and Hayek, were after a theory of institutions, or so he claimed, but they were ill-equipped to study the incentive properties related to the communication of information. In 1972, Hurwicz managed to crack the case; now we understand moral hazard. Possibly significant, his insight was profound, but the contribution apparently did not merit much further elaboration on his part. It is revealing that in Myerson's entry on "mechanism design" written for the *New Palgrave Dictionary of*

Economics (2008), Hurwicz's name appears only in the bibliography, as co-editor of a volume including one of Myerson's papers (the works of Maskin and Myerson are cited several times).[21] When it comes to Hurwicz's role, one feels as if encountering a declassified and heavily redacted embassy wire: two or three sentences in a sea of black.

In pointing out these omissions concerning Hurwicz's career, our intention is not to suggest that he has received inadequate credit—an absurd claim in light of his receipt of the BOS Prize. Instead, our purpose is to illustrate that in their single-minded attempts to find precursors to one specific favorite research program, such authors manage to miss almost everything potentially illuminating about the career of Hurwicz, *even as it pertains to mechanism design*.

To give a sense of what is missed by such accounts, we will present our own brief account of Hurwicz's career, unexpurgated, and truer to the historical record.

Leonid Hurwicz was born in Moscow in 1917, spending most of his childhood in Poland after his family fled Russia following the victory of the Bolsheviks. He was a voracious student, and his training was diverse: he studied law at the Józef Piłsudski University (now the University of Warsaw), but he also studied at the Institute for Experimental Physics. His intention was to follow his father in practicing law, but the mounting Nazi threat caused him in 1938 to depart for London. Then, after the British refused to renew his visa, he went to Geneva, and finally in 1940 to the United States, where he was taken in by cousins living in Chicago. Following the entry of the United States into World War II, he landed a job teaching statistics, mathematics, and physics to army and navy recruits at the Institute for Meteorology. Soon thereafter, on the strength of his statistics background, he was recruited to the ground zero of information processing in neoclassical economics, the Cowles Commission.

In some ways, Hurwicz was a typical Cowles Commission member: an immigrant (Polish, like Lange), skilled in mathematics, with a background in the natural sciences. Among those at Cowles, Hurwicz's experiences specifically attuned him to Hayek's claims about the informational advantages of markets. While in Europe, Hurwicz encountered Hayek (in London) and Ludwig von Mises (in Geneva). Around that time he became an adherent of market socialism. As stressed in chapter 7, at Cowles, Hurwicz reconceptualized an economic system as a kind of communications device, vaguely like a computer, whose properties could be studied to see whether it ran correctly (for Cowles, "computing" a Pareto optimal allocation), possibly as a prelude to technological augmentation. The term he eventually coined for such studies was "mechanism design," although "design" here assumed a very peculiar sense, since Hurwicz neither built anything nor seriously proposed how one could use his "designs" to do so. What, then, were these "mechanisms?" At base they were directions on how to structure communications processes within an economy/institution/organization. But they were lightly sketched, and often given slim motivation: some seemed offered almost in the spirit of a joke.[22] He did propose restricting consideration to processes that were "decentralized," seemingly a concession to Hayek, although not much of one, since it basically left the planning ambition of market socialists untouched.

The fact that he viewed decentralization as a constraint dictated by limitations on the capacity of economic communications pointed to one additional inspiration for his "mechanisms"—cybernetics. Among a Cowles group deeply influenced by cybernetics, Hurwicz was perhaps the biggest enthusiast, claiming that it informed his explorations into decentralized mechanisms, and going so far as to suggest that it could settle Cowles's intractable dispute with the institutionalists. Hurwicz also enjoyed a reputation as a cybernetics expert within the corridors of power of the defense establishment; in the 1960s, he served on the Cybernetics Panel of President Kennedy's Scientific Advisory Committee, which studied the purported "cybernetics gap" between the United States and Russia (and found the state of Russian cybernetics wanting).[23] Hence, the point of mechanism design, at least initially, was to use the imagery of cybernetics to build institutions that could work around the kinds of problems with central planning that Hayek had so strenuously objected to.[24]

Owing in part to the failure to make appreciable headway on "decentralization," not to mention the accumulation of negative results regarding generalized uniqueness and stability of equilibrium by the Walrasians, and to external changes in conceptions of information, by the 1970s, Hurwicz had shifted focus to "incentive compatibility." He claimed, "[A]t some point I decided that since I know people are not angels, perhaps I should not completely ignore the incentive aspect."[25] That the term had appeared in his very first (1955) paper on mechanism design indicates he had earlier considered this possibility.[26] In retrospect, Hurwicz tended to portray the shift as pertaining to the recalibrated topic of study, from informational aspects to incentive aspects of mechanisms. In fact, both sets of work addressed information; *what had noticeably changed was the operant notion of information.*

Informationally decentralized mechanisms were reconceptualized as noncooperative games, with the planner now tasked as designing "incentive compatible" systems for encouraging "truth telling," and the likelihood of such truth telling ascertained by application of the Nash solution concept. The task was no longer to explore limits in channel capacity but, instead, to purportedly force agents to reveal their privately held information. Hurwicz seemed oblivious that he was shifting from the sphere of mere communication technologies to thornier questions of the nature of truth.

Initiating this shift in how to conceive of information almost immediately led to another, where now information would be conceived more in the tradition of Blackwell's idiom. Most of this work was undertaken by students of Hurwicz, or those encouraged by him, often at Purdue, Northwestern, or a long-running series of annual conferences on decentralization sponsored by the U.S. National Science Foundation.

Hurwicz did evince an awareness of computational issues, the third class of information models, but tended to suppress them or otherwise relegate them to the secondary status of "transactions costs."[27] He avoided the more fundamental issues of computability: "feasibility of the computations required by the process, as distinct from possession, transmission, or perception of information is ignored."[28] The task of addressing computability effectively devolved to Stanley Reiter (Hurwicz served on his dissertation committee) and Kenneth Mount. The effect was to lead them further and further from the orthodox Walrasian account.[29]

One therefore observes in the trajectory of the career of Hurwicz a traverse through the various approaches to information available to orthodox economists, covered in chapter 8. Unfortunately, Hurwicz was not explicit in acknowledging his own shifting approach to information; sometimes he acted as willing participant in an effort to marginalize his previous enthusiasms,[30] even as he did not entirely relinquish them in his work. If there was continuity, it lay in what Hurwicz opposed. In ringing these changes, he would feel compelled, time and again, to invoke the name of Hayek, who was a formative influence:

> Much of my own work since the 1950s has been focused on issues in welfare economics viewed from an informational perspective. The ideas of Hayek (whose classes at the London School of Economics I attended during the academic year

1938-39) have played a major role in influencing my thinking and have been so acknowledged. But my ideas have also been influenced by Oskar Lange (University of Chicago, 1940-42), as well as by Ludwig von Mises in whose Geneva seminar I took part during 1938-39.[31]

Throughout his long career in mechanism design, Hayek's challenge remained at the forefront of Hurwicz's mind. In his first work on mechanism design, Hurwicz freely cited the work of Hayek as a primary motivation: Hayek's "The Use of Knowledge in Society" was one of only two cited references in the bibliography, and he framed the entire discussion as a meditation on the nature and significance of the argument advanced by Hayek. He noted, "It was Hayek who indicated some of the desirable features of the manner in which the competitive adjustment process handles the relevant information."[32] At that time, he interpreted the Hayekian doctrine in conformity with the earlier version postulating information as dispersed, but still accessible, provided one employed a method of communication akin to—though better than—The Market. Hurwicz's shift to game theory was accompanied by a changed view of the economic significance of information. Now the focus would be on devising schemes to encourage "truth telling," informed by the belief that economic knowledge was mostly inaccessible, and needed to be more coercively extricated through the market.

In this shift in epistemics, Hurwicz followed his former teacher but political opponent, albeit with a bit of a lag. The final leap would come mostly with the passing of the baton to his students, who would begin to study the computational properties of economic systems, and shifting the emphasis from "mechanism design" to the engineering profession of "market design" for fun and profit. The pirouette from fantasy to fabrication happened haphazardly, but then, with greater conviction.

MARKET DESIGN TAKES OFF

What would market design look like in action? The Great Transformation of economists into engineers did not happen all at once. Let us consider what was arguably the first attempt to "go the market one better"—the design of a market to improve the flow of commercial airport traffic in high-volume U.S. airports.[33] Although the study, which was authored by David Grether, Mark Isaac, and Charles Plott, was originally intended for an audience of officials at the Civil Aeronautics Board and written accordingly, it circulated widely and left a deep impression on academic economists, particularly those cadres most responsible for the development of market design. It attracted numerous citations—including, significantly, in the Markets and Organizations report, as an example of a promising new development in economics. That the study became so influential is made all the more astounding by the fact that it was not formally published for a full decade after its completion, when it was reissued as an "Underground Classic in Economics."[34]

The circumstance that conjured this early instance of market design was the deregulatory wave that had begun to wash across the U.S. political landscape in the 1970s. Airline regulation was an early and conspicuous target: Senator Ted Kennedy held hearings on the performance of the Civil Aeronautics Board (CAB) in 1974; a few years later, in 1977, President Jimmy Carter appointed Alfred Kahn to the CAB. Kahn appointed the law and economics scholar Michael Levine, first as director of the Bureau of Pricing and Domestic Aviation, and later as general director of International and Domestic Aviation.[35] Characteristic of the law and economics movement, Levine advocated for deregulation; Kahn instructed him to set to the task.

At that time, high-traffic airports assigned landing slots via a committee composed of members representing the airlines; the potentially anti-competitive nature of this arrangement had begun to attract scrutiny, particularly since the 1978 passage of the Airline Deregulation Act encouraged entry into the industry, suggesting to many that scheduling pressures would continue to mount. As it happened, Levine had recently completed three papers on agenda influence in committees with Plott, who was a former colleague of his at Caltech; soon Plott, along with his colleague at Caltech David Grether and graduate student Mark Isaac, was invited to submit a report to the CAB on the performance of these committees.[36]

Caltech had by that time established itself as one of the two most important centers of experimental economics in the world (the other being Vernon Smith's University of Arizona economics department). This was due largely to the efforts of Plott, who conducted the first economic experiments there in the early 1970s and immediately attracted an impressive cohort, including John Ferejohn, Morris Fiorina, David Grether, and Roger Noll, along with Levine, as well as William Riker (a visitor to Caltech during 1974).[37]

The marked presence of political scientists and legal scholars in this group provides some indication of the distinct provenance of Plott's approach: though he took inspiration from Vernon Smith— and like him, taught at Purdue when it was a center for organization studies—unlike Smith, Plott arrived at experimental economics via public choice, a path he embarked upon as a student of MPS member James Buchanan at the University of Virginia.[38] The "experimental public choice" that he developed during the early 1970s addressed committees, agenda influence, voting, and negotiation— topics of obvious interest to political scientists.[39] Soon thereafter, this Caltech approach merged with Smith's studies of markets,[40] a development that led Caltech experimentalists to claim expertise in the evaluation of every imaginable kind of decision-making process—public, private, market, voting, bureaucracy, and negotiation, including types that lived only in the imagination:

> Policy analysis is ... removed from preoccupation with existing institutions. It becomes a type of "institutional engineering" whereby the basic principles are used to construct "new" or "synthetic" institutions ... [which] may or may not resemble any existing institutions.[41]

Of course, since such "synthetic institutions" would have had no track record, the ability to generate experimental evidence assumed particular salience. While there was no a priori reason why one class of organizations would necessarily be preferred to another—in principle, a market could substitute for a bureaucratic process and vice versa—in practice, markets would always be the most favored organization, a bias that reflected the Caltech program's neoliberal origins.

Given that the 1978 Airline Deregulation Act called for phasing out the CAB by 1983, finding a substitute for bureaucratic decision making was at the forefront of CAB officials' minds. Plott, along with Grether and Isaac (who together formed a company, Polinomics, and submitted under this name), responded with a proposal to do away with the committees and instead auction off the slots to the highest bidders. It did *not* merely call for establishing property rights in slots and allowing their cash sales (which had been prohibited by the FAA). Instead, the initial assignment of slots would be a highly structured affair, with each airport periodically conducting a formal auction for them.

Polinomics proposed a heretofore-unimagined market, referred to by one reviewer as an "almost Vickrey" auction: it would call for sealed bids, award slots to the highest bidders, and charge the awardees a uniform price equal to lowest winning bid.[42] Although they did acknowledge that complementarities and indivisibilities might compromise the effectiveness of their auction (acquiring a departure slot would be useless without a corresponding landing slot at the destination, at the right time), they offered to remedy the problem by running a continuous aftermarket trading scheme (now, organized as an "open book" auction).

In laboratory experiments, those with the highest induced values tended to acquire the slots; on this basis, the Polinomics report claimed it was a demonstrably "efficient" method. But crucially from the standpoint of the CAB (as well as the airlines, which were deeply involved in the rulemaking process), Plott and his co-authors did not confine their analysis to the pursuit of static efficiency; they insisted they could engineer the market to deliver on a variety of additional policy goals: service to small communities, responsiveness to changing circumstances, safeguarding against monopoly and collusion, promoting long-run industry growth, and increasing airport capacity.[43]

This early example offers us a glimpse of what made market design so novel. Clearly, Plott and his co-authors were suggesting that markets could improve on bureaucratic decision making, a common neoliberal theme at the time. Implicit in their proposal, however, was the position that the way trade was structured would influence the "effectiveness" of the market. This opened up a potential for the economist to apply the lessons of economics to auction design, yet the relationship with previous work in economic theory was strained and awkward. The resulting markets may have borne some resemblance to auction forms previously analyzed in the literature, but the resemblance was usually vague, and the form settled upon would often fail to heed what was considered a central theoretical lesson found in the literature (e.g., the airport-slot auction was *almost* Vickrey). Sometimes this departure would be chalked up to the necessity to recognize political realities; more often there was a solid *economic* rationale: market designers often lacked confidence in extant theories of agency (was meeting criteria of "incentive compatibility" *really* that important for performance?).

Newly developed methods of conducting computerized laboratory experiments meant that market designers could "test" a newfangled market's performance prior to implementing it; one might even explore alternative configurations in an unstructured way. Of course, adjudging performance often meant collapsing a variety of potentially incompatible goals into a single number—"allocative efficiency" or "system surplus"—but the drawbacks were minor when compared with the benefit of rhetorical force: policymakers had neither time nor patience for optimality proofs, but readily agreed that "97 percent efficiency" seemed pretty good, and certainly was better than 85 percent. Hence, one important innovation of the market designers was their development of techniques persuasive to *non-economists*, opening new forms of engagement with the neoliberal state.

As for fellow economists, subsequent interest in this case zeroed in on potential problems with the aftermarket; emphasis on post-market correction seemed only to underscore the drawbacks of the "almost Vickrey" auction. Instead, Stephen Rassenti, Vernon Smith, and Robert Bulfin proposed integrating the individual airport auctions into a single, grand auction in which it would now become possible to submit "contingent" bids for combinations of slots.[44] The resulting maximization program would now face an explosive number of constraints, raising the issue of combinatorial complexity. Rassenti addressed this issue algorithmically, via the instrumentality of the physical computer. Hence, what began as a commission to assist in deregulation rapidly became an elaborate task in management of *information processing*.[45]

Here, the shape of things to come clicked into place: information processing in the service of neoliberal politics, and endorsed by those (authors of the Markets and Organization report) who originally took their charge as rebutting Hayek's position in the Socialist Calculation Controversy. As information was reified, economists began to fix their focus upon organizations, and then markets. The era of market design was at hand.

[10]
THE HISTORY OF MARKETS AND THE THEORY OF MARKET DESIGN

Before 1980, many people believed that The Market was something that has always existed in a quasi-natural state, much like gravity. It seemed to enjoy a material omnipresence, sharing many characteristics of the forces of nature, warranting a science of its own. That science was first called "political economy," and then, after roughly 1870, "economics." The modern orthodoxy of that science, the neoclassical tradition, has always taken the nature of The Market as the central province of economics, has it not?

Assuming so would be premature, as some high-profile orthodox economists have noted: "It is a peculiar fact that the literature on economics . . . contains so little discussion of the central institution that underlies neoclassical economics—the market."[1] And, "Although economists claim to study the market, in modern economic theory the market itself has even a more shadowy role than the firm."[2] Arrow and Hahn's General Competitive Analysis asserts in passing that it takes the "existence of markets . . . for granted."[3] In fact, a judicious and unbiased overview of the history of the first century of neoclassical economics would confirm that its adherents had been much more fascinated with the status and nature of *agents* than with the structure and composition of markets.

Most of the time, the concept of The Market was offhandedly treated as a general synonym for the phenomenon of exchange itself, and hence rendered effectively redundant.[4] Even in the few instances when key thinkers in the tradition felt they should discuss the actual sequence of bids and asks in their models of trade—say, for instance, Leon Walras with his tâtonnement, or Edgeworth with his recontracting process—what jumps out is that they bore little or no relationship to the operation of any actual contemporary market.[5] Mid-twentieth-century attempts to develop accounts of price dynamics were, if anything, even further removed from the increasingly sophisticated diversity of market formats and structures, as well as the actual sequence of tasks that markets accomplish. Spectral auctioneers, mechanical differential equations written in terms of economy-wide "excess demands," markup controversies, cobwebs, and the like had nothing to do with the structure of activities of real-world market participants in their diverse market settings.

Any acknowledged differences in market structures where agents congregated would be treated as second-order complications (viz., perfect competition vs. monopoly) or else collapsible to commodity definitions ("the" labor market, "the" fish market); and therefore The Market in neoclassical economics came to be modeled as a relatively homogeneous and undifferentiated entity. Whether justified as mere pragmatic modeling tactic (for reasons of mathematical tractability) or a deeper symmetry bound up with the very notion of the possibility of existence of "laws of economics," market diversity was effectively suppressed, as one can still observe from modern microeconomics textbooks.

A whole slew of events intervened to undermine this monolithic view of the nature of a generic Market—a full roster that deserves its own historian. Here, we shall have to rest content with a mere incomplete list. The first and foremost catalyst was the transformation of the computer from an icon of a giant calculator, to the increasing appreciation of computers as distributed all-purpose communication devices, starting in the 1970s, and culminating with the spread of the Internet in the 1980s and 1990s.[6] The second was a number of interventions made by regulators from the 1960s onward to improve or otherwise reconfigure specific markets. Here, we point to the work of a number of economic sociologists, such as Greta Krippner, Sarah Quinn, Juan Pablo Pardo-Guerra, Carolyn Sissoko, Martha Poon, Yuval Millo, and a host of others.[7] Their shared perspective is that many dramatic alterations in market formats began as well-meaning "reforms" reacting to short-term political controversies, often without any intention of producing the later upheavals in actual market functions that ensued.

To give a brief flavor of this work, we can point to the exemplary text of Sarah Quinn on the birth of securitization of mortgage debt. She documents the efforts of Fannie Mae to encourage a wider and deeper national mortgage market up until 1968, when Lyndon Johnson sought to hide his ballooning budget deficits linked to the Vietnam War by privatizing Fannie Mae and placing its activities "off-book"—laying the foundation for issuance of standardized mortgage-backed securities, authorizing Fannie Mae to issue securitized bonds based on pools of sanctioned mortgages and sweetened with a number of partially hidden government guarantees. In 1970, Freddie Mac was founded on this same model.

With further amendments that cannot be covered here, the government effectively removed much housing subsidy off its books in the short term, and simultaneously encouraged further "financial innovation" on the part of private issuers of debt. An excess of ingenuity spawned new market formats with new "products." New instruments begat new markets, which eventually came to grief in the global financial crisis that began in late 2007. Similar stories have been told concerning the post-1970 fragmentation of share markets, the rise of credit scoring, special investment vehicles (SIVs), and much else besides.

The third set of events driving intellectual recognition of market reorganization was the growing political strength of the Mont Pèlerin Society and the neoliberal thought collective. Throughout the various histories of market innovations, one observes from the 1980s onward that neoliberal politicians and think tanks enter into policy debates to recast many of the innovations initiated by government regulators as needing to take into account that markets were better judges of the success or failure of those very same market reconfigurations; often they pushed further market engineering under the rubrics of "deregulation" or "privatization." The worship of innovation had come to be trained on markets themselves. One observes this in the blitz to locate financial derivatives in arenas beyond the reach of regulators, or in the demutualization of previously protected stock exchanges.

Economic sociologists have rightly insisted that these battles were over the very definition of what a legitimate market was or should be; more often than not, ideals of efficiency were recast as appeals to *informational efficiency*, such that the market could easily outperform the surveillance of any potential regulator. Thus, actual market structures were being nudged to look more like information processors, extolled by neoliberal politicians and inspired by the computational developments just mentioned.

So markets were changing, and eventually the economics profession had to sit up and take notice. The multiplicity of markets started to break through their academic consciousness in all sorts of ways: the rise of a burgeoning literature on the "efficient markets hypothesis" in finance; the materialization of a dedicated field concerned with "market microstructures"; the coding of market protocols in experimental economics. The proliferation of market minutiae posed a challenge for neoclassical economics: one could continue to merely describe and taxonomize the efflorescence of structures, or else one could grasp the nettle and begin to claim expertise in building some of these novel structures. That latter option was the beginning of the field of "market design."

Consequently, since roughly 1980, the profession converged upon a more "constructivist" approach to markets in the sense that it has become possible, for the first time, to acknowledge that market formats do indeed differ in significant ways; furthermore, it might be possible for economists to intervene in the setup and maintenance of these diverse structures. Where economists once placidly contemplated markets from without, situated in a space detached from their subject matter, so to speak, now they are much less disciplined about their doctrines concerning the nature of economic agency, and much more inclined to be found down in the trenches with other participants, engaged in making markets.

Economists were not the only social scientists caught up in these trends. Science studies scholars and economic sociologists—groups nominally hostile to the economics orthodoxy—soon took notice that economists had adopted a more hands-on approach, and they sought to interpret this as leaving an opening for their own theoretical predilections and potential interventions. Michel Callon has famously argued that "the role of the sociology of economics and the anthropology of economics is precisely to design tools" in order to "influence or structure institutions."[8] His followers came to believe that their discovery of the active intervention of economists in (some) markets—a phenomenon they wish to characterize as economists "performing" the economy—is a major validation of the ontological theses for which they have become famous.

We have argued elsewhere that little solid in the way of usable analysis for science studies and economic sociology can be derived from the performativity thesis.[9] Although this is not the place to rehearse that argument, we would reiterate that by focusing exclusively on the interventions of economists, performativity accounts miss some of the most important reasons why markets are changing. Economists have not wrought their designs solely on their own terms, as we shall argue later in this volume.

An additional problem is that the performativity thesis has begun to influence the way the history of economic thought is currently being written. Here, we point to Ivan Boldyrev and Alexey Ushakov, who credit this work on performativity with inspiring them to draw attention to the "constructive" elements of economic models, and to present as their exhibit one Leonid Hurwicz's practice of mechanism design.[10] According to their performativity-inspired reading, Hurwicz and those influenced by him sought to create algorithms for the purpose of implementing market socialism.

Obviously they weren't able to do so, but once they began to account for incentive structures and experimental results, scaled down their ambitions, and set to the "local" task of designing new kinds of markets and auctions, they did achieve some success. While this necessitated that mechanism design take on board some lessons of game theory and experimental economics, doing so required the program only to add a constructive element, instead of repudiating the prior core commitments of their research program: "Planning became more local, but the principles of mechanism design ... remained the same." Hence, for Boldyrev and Ushakov, developments in economics—such as Walrasian mechanism design, Bayes-Nash game theory, and experimental economics—are understood jointly as complementary activities in the task of building reliable "economic machines."

Here, we find the telltale pathologies of performativity-inspired accounts: the deployment of a new armamentarium of jargon that serves only to gussy-up orthodox textbook accounts of the historical development of economics. There are no ruptures, and nothing is repudiated: all economists are enlisted (team-like?) in the collective project of designing reliable machines. While these kinds of accounts admit many objections, it is the penchant of the performativity-inspired account to collapse fundamentally different and even contradictory research programs into a uniform "economics" that comes to be "performed," which for the historian constitutes its most damaging legacy.

In this case, the misunderstanding extends not only to the most important elements of game theoretic and experimentalist approaches to market design but also to Hurwicz's own project, as we will argue in the following section. True, there may have been a constructivism inherent in Hurwicz's version of "mechanism design," but it was repressed and, crucially, his followers did not care to pursue that avenue. Hurwicz never supervised the building of anything. We noted in chapter 9 that during the 1980s, various and sundry economists began to apply the term "design" to their activities with increasing frequency. But one should not confuse resort to the same idiomatic term with overt agreement on the appropriate activity of the economist; in their hands, "design" referred to an amazingly wide variety of activities (what, *precisely*, was being designed?), among which only a limited subset would involve the construction of a concrete institution in the world. Within the profession, the opacity of this term has led to all matter of confusion about the conceptual relationships between "mechanism design," "market design," "auction theory," and game theory, as we will argue below. Unfortunately, performativity-inspired accounts both reproduce and amplify this confusion.

The crux of this problem, we insist, is a failure to pay due attention to what information has meant to these economists. We devote the remainder of this book to remedying this confusion.

THE NARRATIVE TRAJECTORY OF THE HISTORY

Our plot for the remainder of this book consists of two major components: (1) an abstract intellectual trajectory traced by the economics profession concerning the nature of markets, built up from elements of the previous chapters; and (2), a history we hinted at above, which has begun to be written outside the economics profession of late, that describes how *actual markets* have been profoundly altered in the late twentieth century, often (but not invariably) with the participation of economists. Both components exist to drive home the lessons from our history: (1) there is *still* no single orthodox economics of information, even at this late date, although there are *trends* that can be described, if not completely explained; and (2) markets in the twenty-first century are continually being reengineered to look more like information processors, even if they didn't start out that way. This time things really are different.

The career of Leonid Hurwicz, sketched in the previous chapter, illustrates the intellectual trajectory. Neoclassical economists, initially located at Cowles, appropriated formal models of information from the sciences. These models served to structure distinct and implicitly incommensurate research programs within the orthodoxy. Initially, after the fashion of Claude Shannon, economists began to treat information as a thing-like commodity. Some quickly relinquished this view of information, while others such as Leonid Hurwicz pursued this will-o-the-wisp until it could no longer be maintained.

After mostly giving up on this option, a broad range of economists looked into treating information as statistical induction. Information in these models was envisaged as the output of inductive inference, usually Bayesian, structured by Blackwell's formalism, where information was conceived as measures over partitions of states of the world. This version found a snug home in game theory and macroeconomics. In a separate development, fewer economists explored the notion of information as symbolic computation, inspired by studies of Turing Machines, leading directly to computerized smart markets. These three methods of modeling information have constituted the orthodoxy's main lines of development. Other approaches were possible (Herbert Simon's simulations, for instance), but never caught on.

At the same time, the significance of agents' epistemological capacities underwent subtle transformation. In chapter 6, we sought to characterize this by following Friedrich Hayek's ideas about agents' knowledge across three phases of his career. First, Hayek portrayed knowledge as dispersed, and in some sense "local"—very difficult to gather, but not impossible to do so. This was the Hayek of the "Use of Knowledge in Society," the most oft-cited of Hayek's works. Methods of aggregating up individual agents' knowledge would have a crucial bearing on the operation of the well-functioning economy. During the second phase, Hayek revised his treatment of knowledge, now portraying knowledge as "tacit" and inaccessible to self-examination. *The Sensory Order*, representing this phase, marked Hayek's first foray into the study of human cognition. Finally, Hayek came to portray knowledge as completely disengaged from the consciousness of the knower. This was the Hayek of "Competition as a Discovery Procedure," wherein he deemed much of agents' conscious knowledge as irrelevant to the operation of the well-functioning economy. In this incarnation, some knowledge could only be discovered by the market, and so in this final phase Hayek conceived ideal intentionality of individuals as acquiescing in the market's signals.

One primary reason for our detailed examination of the work of Hayek was that, unlike that of the orthodoxy, which evinced a tendency to uncritically conflate information with knowledge, Hayek's work more readily lent itself to understanding the ways different formal models of information related to different roles of *knowledge* in the economy. Members of the orthodoxy did not often find occasions to have much to say on this matter. Nevertheless, our primary argument for the remainder of this chapter is that bringing this relationship to the forefront is uniquely illuminating. In a way which might appeal to economists, we will lay out the various possibilities in a graph, which imagines the possibilities as if they were situated in an abstract space of "informations."

We propose to do so by erecting a device we call an "Information Space," which we illustrate in figure 10.1. Arrayed up the vertical axis of this Information Space, one finds the three modeling strategies for depicting information: information portrayed as a thing, as Blackwell induction, and as symbolic computation. Along the horizontal axis, we encounter diverse attitudes about the role of

Information Modeled As:

Computation

Blackwell Induction

Thing

Knowledge Matters | Knowledge Inaccessible/Tacit | Cognition Irrelevant

Significance of Agents' Knowledge:

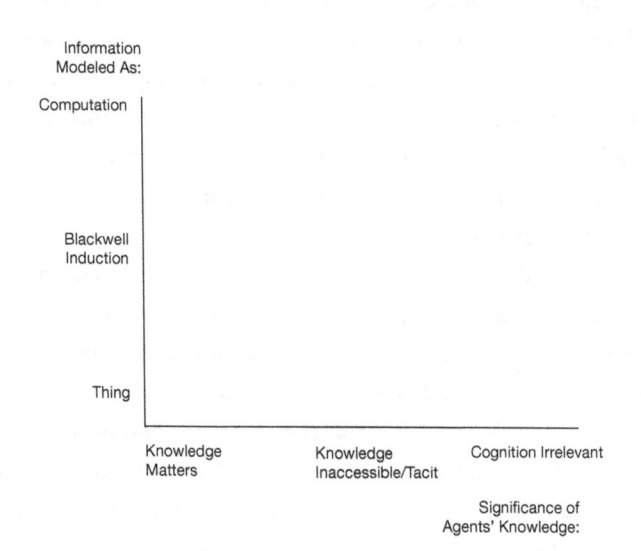

Figure 10.1. Orthodox Trajectory Through Information Space, I.

knowledge in the operation of the well-functioning market: an epistemic attitude that agents' conscious knowledge matters; or is inaccessible or tacit in some or all respects; and the precept that individual cognition is effectively irrelevant to the market's operation.

We situate our history in this Information Space to help isolate the most significant aspects of the intellectual dynamic within which the orthodoxy finds itself situated, and to capture the options introduced in earlier chapters. Interactions between Hayek and the orthodoxy drew our attention to important issues of agent epistemology, which have often escaped notice. Hayek influenced the activities of many at Cowles, and the Cowlesmen acknowledged his influence. We illustrated this most vividly for the case of Hurwicz, but as we will see later, nearly every notable economist concerned with information felt compelled to respond to Hayek in one way or another. The range of attitudes held by Hayek toward knowledge also assumed significance for structuring the activities of the orthodoxy: these are topics encountered in the following chapters.

Moreover, the diagram helps us to recognize that individual economists are not uniformly distributed throughout this space. Rather, a given economist's position along the first dimension has tended to be coupled to his position along the second dimension; "off-diagonal" information–knowledge combinations, while not impossible, have proven unstable.

The centrality of information to the economy has become a pervasive theme in modern economics, but it turns out there is one subset of modern economics that should play a particularly significant role in any intellectual history of the economics profession. This is the shift within the modern profession from the description of markets "from the outside," as it were, to participation in the design and implementation of markets as hands-on engineers of the economy. It should be obvious that this is an epoch-making departure in the history of the praxis of economics, but also, it is an irreversible change in the stance of the economist toward the agents that populate his models. Hence, issues of agent epistemology actually rise to the top of concern when economists claim to be able to go one better than existing markets. Ground zero in this Great Transformation has been the theoretical tradition of so-called market design.

Acknowledgments that market design theory and practice provide the exemplary instance of modern information economics are commonplace: one survey of the field pronounces Vickrey's (1961) game theoretic study of auctions as the *Wealth of Nations* of information economics.[11] Equally significant is the widespread impression that today's markets are "all about data and information." Markets were apparently being reconfigured so the conveyance or masking of information could assume a central role—one thinks here of the securitization of mortgages and the use of markets to allocate wireless communications licenses—forcing market designers to adjust their approaches to claim they were on top of these developments. Markets were becoming something different from what they had been before WWII; and economists would not defer to others to serve as prophets of the new order. Thus, the history of market design constitutes the backbone of any future history of the economics of information, and it constitutes the topic of the rest of our book.

THE SCHOOL FOR SCANDAL, AND THE SCHOOLS OF DESIGN

Since the design turn, economists have enthusiastically trumpeted the accomplishments made possible by it; yet nevertheless, they have never been entirely clear how market design fits into their mental maps of orthodox theory. As of January 2014, the *Journal of Economic Literature* (JEL) designated "Market Design" an official field of study. It now appears as a subcategory under "Market

Structure and Pricing," alongside "Auctions."[12] A closer look at the JEL subject guide yields the following description:

D470 MARKET DESIGN

Guideline: Covers studies concerning the design and evolution of economic institutions, the design of mechanisms for economic transactions (such as price determination and dispute resolution), and the interplay between a market's size and complexity and the information available to agents in such an environment.[13]

The JEL inserts as a caveat, "Purely theoretical studies concerning mechanism design should be classified as D82." D82 is "Mechanism Design," classified as a subtopic of asymmetric and private information, alongside moral hazard, adverse selection, signaling, and principal-agent models. However, these literatures tend to blur the JEL's sharp distinctions. In the introductory chapter to *The Handbook of Market Design*, Alvin Roth, along with his co-editors, writes that market design "applies the formal tools of game theory and mechanism design."[14] But, wait: elsewhere, some describe the game theoretic study of auctions as a "major application of mechanism design theory,"[15] whereas for others mechanism design "builds on the theory of games"[16] or is "a part of game theory."[17] Adding to the confusion is the common practice to refer to auctions as *types* of mechanisms.[18] However difficult it is for the novice to grasp how these topics relate to one another, the experts are apparently faring no better; after posing the question "What is market design?" one well-respected game theorist quipped, "whatever Alvin Roth says it is."[19]

Attending to the specific notion of information used can go some distance in clarifying a confusing intellectual formation. In particular, it may aid in the conceptualization of research "schools." Recall that over time, economists' conceptions of information changed from the "thing" characteristic of Shannon's approach, to an inductive notion of information, characteristic of the approach of Blackwell, and finally to a computability notion of information. Viewing these developments from the vantage point of market design leads us to make our first observation:

Changes in how economists conceptualized information led to changes in how economists conceptualized markets.

During the first period, economists conceived markets generically, undifferentiated, left devoid of any institutional specificity. During the second, economists viewed markets as taking one of a handful of highly stylized formats (the English auction, the Dutch auction), understood in terms of how they might reveal information and assist inference. During the third era of research, individual markets have been viewed as algorithms—and like algorithms may serve a variety of purposes.

Now we can begin to better appreciate why Friedrich Hayek is a major protagonist in our drama. In addition to inspiring the work of Cowles economists such as Leonid Hurwicz, under certain interpretations Hayek and fellow neoliberals stressed the constructed nature of the economic order. Eventually, market designers would come to appreciate the multiplicity of forms that markets could instantiate and the purposes these forms could serve, after the manner of the neoliberals. They would sanction that their understanding of how markets communicate and process information endowed them with an expertise in setting up markets, an expertise linked to their views about the role of agents' knowledge. Although Hayek was far more explicit about this role, notions of knowledge had become significant to the orthodoxy and Hayek alike.

First, economists viewed information as dispersed among individuals who, while not all-knowing, were aware of their own values and used this knowledge in making "local" decisions. Subsequently, economists questioned the nature of the information held by the people involved. Knowledge became seen as increasingly personally inaccessible, eventually entailing that one should view agents as possibly mistaken about their knowledge. Eventually, economists came to more or less disregard what people thought: cognition should not matter to the operation of the market. Inspired by this insight, we make our second observation:

Changes in economists' attitudes toward agents' knowledge brought forth changes in how economists viewed their own roles.

Those who viewed individuals as possessing valuable knowledge about the economy generally conceived of themselves as assisting the government in collecting and utilizing it; those who viewed individuals as mistaken in their knowledge tasked themselves as assisting participants in inferring true knowledge; and finally, those who viewed people's knowledge as irrelevant to the operation of markets tended to focus on building boutique markets.

By linking the knowledge presuppositions of the orthodoxy to the sequence of views adopted by Hayek, we illuminate a heretofore-unappreciated feature of modern economic design: over time, economists have relinquished a concern for ensuring that markets give people what they want, and increasingly insist that they can make markets produce any desired outcome *regardless of what people want*.

We now return to the Information Space, and augment it by relabeling the "Information" axis as "Information-Markets" and the "Knowledge" axis as "Knowledge-Roles," as shown in figure 10.2. The coupling between axes is preserved: following a diagonal ray emanating from the origin, we encounter, in order, three schools of design: the Walrasian mechanism designers, the Bayes-Nash

THE HISTORY OF MARKETS AND THEORY OF MARKET DESIGN

```
Information-
Markets:

Computation         |                        Experimentalist
                    |                        School
                    |
                    |
Blackwell           |   Bayes-Nash
Induction           |   School
                    |
                    |
                    |
Thing               |   Walrasian
                    |   School
                    |_____
                    Knowledge      Knowledge           Cognition Irrelevant
                    Matters        Inaccessible/Tacit
                                                   Knowledge-Roles:
```

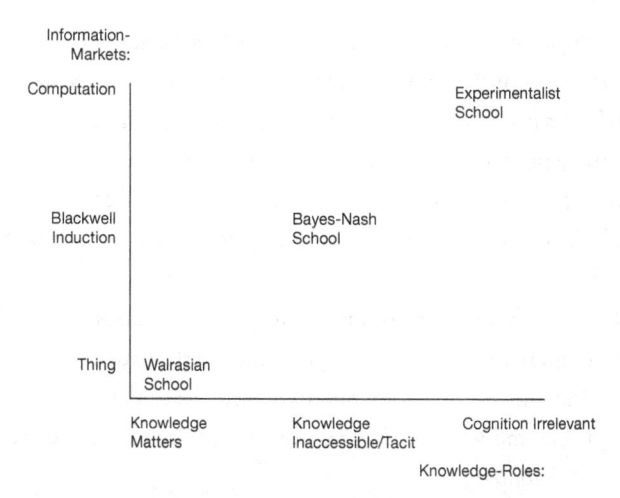

Figure 10.2. Orthodox Trajectory Through Information Space, II.

game theorists, and the Experimentalists. Moving outward from the origin is akin to moving through time in the history of market design, in the sense that the Walrasian School was initiated first (in the 1950s), the Bayes-Nash School next (in the 1960s), and the Experimentalist School last (in the 1970s).[20] While there is nothing necessary about these pairings, in a purely logical sense, and while one or another economist can be found at some point in time in "off-diagonal" space, the pairings have proved to be highly compatible, for reasons we will elaborate below.

This compatibility of theoretical positions on information, markets, and knowledge, culminating in variations in the economist's role in society, is our main rationale for using the term "School," even while members rarely apply the term to themselves (unlike, say, the Chicago School of Economics). However, as we will see in chapters 12 and 13, it is relatively common for members of the Bayes-Nash and Experimentalist Schools to draw distinctions between their research programs.

In the following chapters, we will flesh out a narrative of the trajectory of the profession through this Information Space. It will not qualify as a progress narrative, but it will follow a narrative chronology, and a pronounced tendency. In our chronology, the orthodoxy begins somewhere near the origin and moves toward the "northeast." We will suggest four possible reasons for this movement.

1. Economists have responded over time to successive developments of the notion of information in the natural sciences. The movement from Shannon information to computational models in collateral sciences itself imposes one sort of chronology.
2. Economists have progressively moved away from pure agent-conscious self-awareness as a nonnegotiable desideratum of economic models. This is an epoch-making transformation, and bears profound consequences. The premier conundrum involves the issue of whether agents should be portrayed as being "smarter" than the economist, or fully aware of their own cognitive abilities.
3. The politics of the profession have become increasingly neoliberal, perhaps in step with the larger populace.
4. There has been a separate historical trend of *markets in the world* being reconstructed as information processors.

No one before us has thought to ask how each of these circumstances has revolutionized the economics profession. Previous histories instead displayed a penchant to overwhelmingly focus on evolving notions of disembodied agent rationality (viz., the adaptive agent, the rational expectations agent, the all-too-human agent of the behavioral economists), but clearly one cannot specify the status of rationality without a concomitant commitment to what knowledge is—or, indeed, where it is located.

[11]
THE WALRASIAN SCHOOL OF DESIGN

As we previously argued, Hayek's insistence on the economic importance of information in the Socialist Calculation Controversy piqued the interest of several economists centered at the Cowles Commission, and these economists responded by devising frameworks for treating information transfer within economic systems. Among these new "information economists," Leonid Hurwicz's responses assumed signal importance. Today, again, Hurwicz is known primarily for winning the 2007 Bank of Sweden Prize for enhancing understanding of how markets work, in view of the fact that "information about individual preferences and available production technologies is usually dispersed among many actors who may use their private information to further their own interests."[1] But he assumes heightened importance for our narrative for the early models that he and his students built concerning the gathering of information in economies.

Hayek and market socialists alike initially conceived of information as dispersed throughout the economy, essentially lodged within the brains of agents populating it. Consequently, Hurwicz provided one of the primary market socialist responses to Hayek: the pixie dust could be reconstituted. In time-honored American fashion, the solution consisted of merely in coming up with the right technology. For Hayek, The Market was a "vast telecommunications system," and so it would be for Hurwicz. But Hurwicz's added twist was to focus his redemptive concern on "mechanisms"—generic information-gathering processes patterned on the tâtonnement, a kind of trial-and-error device devoid of institutional specificity. Because of this resemblance, we call this approach the *Walrasian School of Design*.

In retrospect, both Hurwicz and Reiter openly admitted their starting point was Shannon's thing-like version of information:

> In ordinary usage information means knowledge about something. In more technical settings information can also mean anything that reduces uncertainty. Shannon's well-known measure of information was developed to analyze the capacity required to transmit signals that go between a sender and a receiver without regard for the meaning of what is transmitted. Jacob Marschak, among others, sought to apply this concept and measure of information in economic settings, and eventually abandoned that enterprise, because it seemed that the "amount" of information measured by the Shannon formula (entropy) has no necessary relation to the relevance or usefulness or value of the information in economic decision-making.[2]

The central characteristic of the Hurwicz program was to further reify "the message" in hopes of circumventing the obvious drawbacks of the Shannon formula. It was almost as if Cowles economists felt they could plausibly reinvent formal information theory—or at least bend it more to their liking. In this incarnation, messages were still "things" that got passed around in a strange parallel structure, situated beside or superimposed on the conventional Walrasian market. In this spasm of hubris, issues of semantics, cognition, and even syntactics became horribly mangled:

Messages may include, for instance, formal written communications within a firm, such as sales, production or accounting reports. These typically have conventional formats. They usually consist of an array of blank spaces in which numerical (and sometimes alphanumeric) entries are made according to given instructions. Therefore, such a report is an ordered array of variables, whose possible values form a (vector) space.[3]

If this sounds a lot like the punch cards of that bygone era's mainframe computers, then one would not be mistaken. In effect, members of the Walrasian Design School took it upon themselves to speculate about the imaginary operations of imaginary computers, without, of course, being held to much of anything in the way of constraints of computer science or information science or linguistics or human psychology. This license to dream was the primary warrant of their ambition to design an organization from scratch, and maybe even an entire economy.[4]

Additional prominent members of the Walrasian School included Stanley Reiter, Kenneth Mount, John Ledyard, and Hiroaki Osana; one might also add Hurwicz's colleague at Cowles, Jacob Marschak, along with his son, Thomas Marschak of RAND. The Walrasian School sought to devise other such mechanisms that would communicate non-price information in the same rough manner as the tâtonnement itself, with the hope that this would improve upon the performance of The Market. We would also suggest that, although members of the MIT school such as George Akerlof and Joseph Stiglitz almost never dealt with full-blown Walrasian structures in their own models, their treatment of *information* almost exactly conforms to the Hurwicz setup, and therefore, for our purposes, they might be considered honorary members of the Walrasian Design School. Their desire to "improve the world" is a less systematic remnant of their predecessors, the more confident postwar market socialists.

The idiosyncratic move of this school was to simply ignore Hayek's first principle that The Market was by itself a superior information processor, and proceed to imagine other Rube Goldberg information processors in vague, sketchy outline. Essentially, however, they ended up back at the seemingly contradictory point that, in the absence of externalities or other arbitrary glitches but compared to their imaginary zoo of "mechanisms," it was the pure competitive market model that achieved the Pareto optimalities they cherished so highly.[5] Not for the first and only time, the market socialists had seemed to lose track of the crux of the political arguments they had acknowledged as their own deep motivations.

We could summarize the Walrasian School setup using the simplified diagram provided by Reiter in 1977, and reproduced here as figure 11.1.

In the canonical version of the model of communication, the economy is populated by $i = 1,\ldots, n$ participants consisting of firms, households, and a central agency. Somehow, all agents come equipped with an initial endowment of "information," though little or no effort is devoted to explaining how it got there, except for the odd stipulation that "no agent by himself knows enough to figure out the feasible allocations."[6] Their preferences, technologies, and endowments constitute the economic environment e. The universe of all possible "economies," which often reduced to the rather less imposing option of a roster of all possible "trades" given preferences and endowments, was designated by capital E. At each time period $t = 0, 1,\ldots, T$, all n participants emit messages m_t drawn from a mysterious preexistent (usually Euclidean) message space M. Participants robotically select their messages from the rule defined by the difference equation $m^i_{t+1} = f^i_t(m_t, e^i)$. The planning procedures are defined by the calculations, the response functions f^i's, dictated by

THE WALRASIAN SCHOOL OF DESIGN

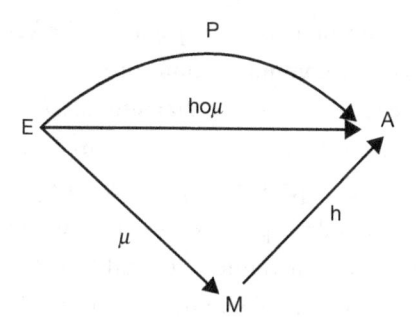

Figure 11.1. Reiter's Schematic.

Reiter, Stanley. 1977. "Information and Performance in the (NEW)² Welfare Economics." *American Economic Review* 67(1): 226–234.

the planner for each participant and the central agency to undertake. The planner seems to double as information auctioneer, imposing a lockstep sequence of information flow prior to physical exchange. The message exchange continues until each participant settles on a stationary or "equilibrium" message, at which time the process ends. Upon completion, the equilibrium message vector m^* is plugged into an outcome rule h (generally ignored when studying the informational properties of the mechanism), yielding directions for commodity flows, $a = h(m^*)$.

See the similarity with the tâtonnement? The mechanical process of passing around the "messages" is summarized through the instrumentality of a "message correspondence" $\mu(e) = \{m \in M \mid f(m,e) - m = 0\}$ identifying the fixed point of message convergence. Needless to say, all markets operated in an identical manner; in parallel, all communication was carried out in the same fashion by all agents. Through the mobilization of information of this sort, information about economies E is translated into final trades in outcome A.

The Walrasian School focused almost entirely on the dimensionality of the equilibrium message vector m^*, which from their perspective would provide a measure of information use. They regarded limitations on channel capacity as the key informational problem, and viewed communications problems "analogously to a limitation of the (cross-sectional) diameter of a pipe restricting the flow of a fluid through that pipe."[7] For example, the tâtonnement process was said to require a message exchange space with the dimension of commodity space because the auctioneer's messages consist of prices for each commodity and the agent's replies consist of quantities for each commodity. Commodity dimensional messages turned out to be a lowest bound for the attainment of Pareto optimal allocations in this setup,[8] which served as a default performance measure. So the Walrasian School of Design sought to identify mechanisms that shared the "informational efficiency" of The Market, which following the Arrow-Debreu results, could also attain Pareto optimality—albeit in only highly circumscribed conditions.[9] Manifestly, they regarded information as a "thing" conveyed through costly information channels, after the manner of their imaginary reconstruction of Claude Shannon's ideas. From a distance, it looked like "information" was being treated with little conceptual distinction from other commodities, so attaching a "price" to it presented itself as a natural extension. Citing such costs, Walrasian mechanism designers declared a preference for "decentralized" mechanisms over "centralized" ones.

The design task for economists under those circumstances amounted to dictating to agents and the planning board the types of calculations they were supposed to undertake and the communications permitted to undertake them. They devoted much effort to identifying "decentralized" economic systems, and distinguishing them from "centralized" ones. But when it came to interpreting the economic significance of such criteria, an identifiable referent for the "cost" of communication was lacking: Did it arise from resources devoted to transmission? Was it a "transaction cost?" Was it a cost of calculation, incurred by agents in the process of thinking?[10] At various times, members of this school would talk this way, referring to the work of Herbert Simon, but such considerations of actual bounds on rationality (or cognitive considerations more generally) never gained a foothold in the analysis of this school. The original slapdash nature of the treatment of cognition could not be hidden by the *post hoc* imposition of actual thinking after the fact.

Curiously, one attempt to do something about this impasse ushered in the work for which Hurwicz is most fondly remembered by the orthodoxy. There was no whiff of game theory in his early work, but as the profession came to embrace Nash equilibrium in the 1970s, Hurwicz decided to go with the flow. In attempting to render the message processing rules f and μ slightly less robotic, Hurwicz hit on the option of further augmenting these processes as the outcome of a game played over message spaces by the agents.[11] This was a rather special sort of game, one where the moves were selection of different messages, rather than the usual array of actions; the question of whether the payoffs were denominated in "knowledge" or something rather more pecuniary got elided. Nevertheless, it was this initially small amendment to the Walrasian design schema that got blown all out of proportion by the next generation, as the great desideratum of "incentive compatibility," and is currently treated as Hurwicz's great contribution to the market design literature.

Perhaps the bulk of the Walrasian design model eventually got neglected simply because it was a machine so awkward and implausible that it was hardly worth defending. The notion of an "auctioneer" preventing everyone from trading until he could converge on a set of zero excess demands was fanciful enough; but the posit of some spectral "thought auctioneer" that prevented everyone *from engaging in any thought or action whatsoever* while they conducted a stylized exchange of messages (sometimes prior to a separate round of conventional tâtonnement) was simply one sloppy contrivance too far, even for the economics profession. MPS member Fritz

Machlup pointed out the strained incongruity, but furthermore linked it to a misunderstanding of the later Friedrich Hayek:

> I still must do some hard thinking on the problem of the "adjustment mechanism." Your idealization of adjustment as a sequence of messages and replies *before* any decision or action is taken seems to me like a nice game but not too helpful in the description and explanation of economic processes. To me, *every* actual price is a disequilibrium price on its way towards a completely fictitious equilibrium that always lies in the future.... The question is how far should we go in our abstractions ... information and informed (or misinformed) decision-making are essential, and the processes of dispersed versus centralized knowledge call for comparison on several scores.[12]

As Hayek had stated repeatedly, the obsession with "equilibrium" distracted attention from the very functions that neoliberals had attributed to the market.

Nevertheless, Hurwicz did advance the orthodoxy in at least one important respect: his reconceptualization of mechanisms as the "unknown of the problem" did make it less acceptable for Walrasians to continue ignoring actual market structures, a necessary precondition for the development of market design, as we noted in chapter 10. Kenneth Mount and Stanley Reiter continued to explore message systems, while at the same time making more explicit the connection of mechanism design to computational themes.[13] But they acknowledged that exploring computational properties required them to relinquish the strict separation of the person from the market; therefore, they had unwittingly demonstrated that it is not possible to consider computational

THE WALRASIAN SCHOOL OF DESIGN

considerations from within the Walrasian program, since this separation was one of the cardinal precepts of the neoclassical school. Those wishing to integrate computation into economic analysis would have to explore another route. Thus, unwittingly, the Walrasian School pioneered the themes that would eventually dictate its own irrelevance.

The Walrasian organon was eventually circumvented by those who wished to grapple with the developments in actual markets themselves. The Walrasian School was far too removed from structural developments in markets to offer any methodology to address them. The task of responding to such real-world developments would be undertaken by the next two schools of design.

[12]

THE BAYES-NASH SCHOOL OF DESIGN

When a graduate student in economics thinks of market design, the first thing that is likely to come to mind is the efforts of game theorists to theorize the incentive properties of various auction forms. One encounters this in core microeconomics courses, as well as in courses more specialized in game theory and even in industrial organization. But here we will resist the temptation to equate the efforts of game theorists with the entirety of market design, stressing instead features specific to this distinct approach.

The historical origin of this school is located in the work of William Vickrey of Columbia University, winner of the 1994 Bank of Sweden Prize, and the namesake of the "Vickrey auction." Vickrey supervised Kenneth Arrow's dissertation, later published as *Social Choice and Individual Values*, and later made his own contributions to the resulting "social choice theory." In a 1960 study, Vickrey noted that "strategic misrepresentation of preferences" might prevent the government from gathering information to construct a social welfare function.[1] A year later, he raised a similar concern with the market socialist proposal of Abba Lerner. In his *Economics of Control*, Lerner had proposed a "counterspeculation" method, to be used by a central board to counteract monopolists' (and monopsonists') price-setting power by estimating and guaranteeing a competitive equilibrium price. In other words, Lerner's analysis had suggested to Vickrey that active efforts might be required to gather diffuse information together in one place. In both papers, Vickrey had expressed a novel concern: economists who had hoped to assist the government in gathering dispersed information would encounter a problem. This problem was, in a word, *mendacity*: those holding the crucial information had the capacity to distort it, and for strategic reasons might be expected to do so.

Vickrey responded to this new problem in a way that will by now seem commonplace: he explored the incentive properties of four auction types—the first-price and second-price sealed bid, the English, and the Dutch auctions—and used Nash game theory to do so. From today's vantage point, it is tempting to become excessively fascinated by Vickrey's appeal to game theory as a generic logic of strategic choice, and consequently to ignore the most significant features of Vickrey's work. For Vickrey's version of epistemology, it was no longer possible to hold tight to one's private information—so long as the analyst crafted a method to get in your head to pry the information out of there.[2] To do so would require employing one of a handful of "incentive compatible" auctions. We know that this idea would eventually be greeted with much fanfare, but at the time pretty much everyone ignored Vickrey's use of game theory; even Vickrey would downplay its significance, as "one of my digressions into abstract economics, at best of minor significance in terms of human welfare."[3]

It would be left to other scholars sharing the Cowles Commission's enthusiasms to make the most significant developments along these lines. The key figure here was Robert Wilson of Stanford's Graduate School of Business who, although not formally affiliated with Cowles or RAND, came to share many of their enthusiasms.[4] Wilson wrote his PhD dissertation on concave programming at the Harvard Business School under the statistical decision theorist Howard Raiffa.[5] That such work would be carried out in the seemingly unlikely location of a business school was due to the recent transformation of a small handful of elite business schools, including Harvard, into "Centers of Excellence" for the study of operations research, supported by the Ford Foundation and the Office of Naval Research, and with the heavy involvement of scholars affiliated with RAND.[6] Raiffa exemplified these trends: his hiring by the Harvard Business School (from Columbia) was financed by a Ford Foundation grant to hire a mathematical statistician[7]; he published several works financed by the ONR, and was sponsored by the Ford Foundation to teach at the Institute for Basic Mathematics for Application to Business (to teach mathematics to professors teaching in business schools).[8] His interest in applying Bayesian decision theory to managerial economics (he would later be credited with inspiring a "Bayesian revival" in the 1960s) led him to form the Decision Under Uncertainty Seminar at Harvard, to which Wilson regularly contributed.[9]

Upon his graduation in 1964, Wilson accepted a position with Stanford's Graduate School of Business, also a designated Center of Excellence, and therefore in the midst of its own transformation. Wilson then turned his attention to the characteristically Cowles-style topic of decentralization under uncertainty and, that same year, encountered John Harsanyi at Jacob Marschak's colloquium at UCLA.[10] Soon thereafter Wilson began to model agents in the context of auctions as engaging Bayesian inference—so soon, in fact, that his first paper on auctions predated the publication of Harsanyi's famous papers on Bayesian games.[11]

Wilson's Stanford department became the first institution devoted to the study of modeling Bayesian actors interacting in markets; along with his students Armando Ortega Reichert, Paul Milgrom, and Peter Cramton, Wilson would form what we call the *Bayes-Nash School of Design*.[12] Later, the center of gravity shifted to

THE BAYES-NASH SCHOOL OF DESIGN

Northwestern University, as the program initiated there by Stanley Reiter came to house an increasing number of game theorists.

Subsequent histories credit Wilson with establishing a new "tradition" in economics. But what were the features of this tradition? Some interpret it as pertaining to understanding "how information is distributed and manipulated, concealed and revealed."[13] Literature surveys and textbook accounts often portray the innovation as picking up on scattered previous insights and providing generalization: Vickrey studied only commodities with "private values," meaning that values for commodities were idiosyncratic and personal; now we can study "common valued" items—goods that possess an objective value, such as an oil tract, or to take a classroom example, a jar of pennies. But approaching matters in this way causes us to miss the most significant point of this later work. The operations researcher Michael Rothkopf was one of the few close observers to put his finger on the true significance of the innovation:

> Quite recently publications have begun to appear that indicate that operations researchers are starting to construct bidding models that are realistic and that consider simultaneously the optimality of the decisions of all bidders. The new factor taken into consideration in these models is the uncertainty faced by the bidders as to the value of the subject of the auction. In most of this work, the uncertainty of each bidder is restricted to the value of the subject of the auction to his competitors. Only Wilson has begun to take account of the uncertainty of a bidder about the value of the subject of the auction to himself.[14]

Agents no longer knew their values—their knowledge could now be *wrong*, and much in need of correction. Economists of the Bayes-Nash tradition would generously offer to help them out.

This correction would take place within the context of an auction game. This school portrays all bidders as viewing the auction game in the same way: the structure of the game is purportedly "common knowledge." Bidding against other bidders immediately raised the specter of having to take competing bidders' beliefs into account, and therefore the complexity of dealing with "beliefs about beliefs." This is where Harsanyi's device of player "types," discussed in chapter 8, makes its appearance: the complex hierarchies of "beliefs about beliefs" are collapsed into a single statistic, dubbed a "type."

Initially, I am presumed to know only my own "type," and will assume that I am the same as my opponents (i.e., we are the same "types"). As information is released over the course of the auction, I come to appreciate how my opponents differ from myself. This appreciation matters not only for strategic reasons but also for epistemic reasons: because the roster of types is presumed to be stochastically distributed around a true-valued mean, it is only by participating in an auction that I come to know my own value. How this works can best be understood by considering a typical model of an "English auction" conducted for a single item for sale. In an English auction, the price of the item for sale starts out low and rises until all bidders drop out save one. The lone bidder remaining "wins" the item, and pays an amount equal to the price prevailing at the time the second-to-last bidder dropped out. According to the Bayes-Nash approach, bidders should use the information released by their competitors dropping out of the auction to reconsider and recalibrate their own valuations, and should continue to bid so long as the expected value of winning the auction conditional upon all remaining bidders dropping out is greater than or equal to the price of the good.[15]

For the purpose of tractable analysis, it is typical to assume the game to be "symmetric," allowing the analyst to proceed by focusing on the strategy of a single bidder. There are $N=\{0,\ldots,n\}$ participants in the auction, bidding for items valued according to the function $u=u^i((x,p),\vec{t})\equiv v^i(x,\vec{t})-p^i$. This states that the individual utility of each participant is determined by the allocation decision arrived at by the auction (x), the price paid for the item (p), and a "type profile" $(\vec{t})=(t^1,\ldots,t^N)$, which includes each participant's type. The analyst evaluates mechanism "performance" by comparing these outcomes to a suitable criterion, sometimes revenue maximization, but often the maximization of the aggregate value of the participants, $\Sigma_{i\in N}v_i(x,\vec{t})$, which is rendered in plain language to mean, "allocate items to those agents who value them the most."[16]

The crux of the design problem stems from the belief that agents suffer from uncertainty regarding their own values, a circumstance made formal by the structure of the valuation function, which incorporates the assumption that each agent's values depend on the types of every single participant. According to the Bayes-Nash school, agents resolve this form of uncertainty by coming to learn the types of other participants, a seemingly impossible task made theoretically possible in part by the assumption that the agents' beliefs are assumed to be derived from a common prior distribution π (i.e., the "Harsanyi doctrine," discussed in chapter 8).

The analysis proceeds by identifying an equilibrium bidding strategy, $\beta^{i*}(\cdot)$. The game theorist's ideal sophisticated agents avoid being disappointed by their (usually incorrect) appraisal of the value of what they are bidding on by adopting a strategy that makes prompt use of all the information that becomes available through the instrumentality of the market. Two kinds of information that may be discerned from the operation of the auction are relevant to this task. By publicizing at each stage the number and identity of the participants remaining, an open auction allows bidders to use this information to recalibrate their reservation values. In equilibrium, bidders should adopt a strategy that successfully revises their beliefs. Formally, $\beta_n(s, p_1, ..., p_n)$ gives the reservation price of a bidder of a type s when n bidders have already dropped out at prices $p_1 \leq \cdots \leq p_n$, and $t^{(n)}$ gives the nth highest type among this bidder's competitors.[17] The result, which turns out to be a (Bayes-Nash) equilibrium bidding strategy, is worth examining in detail.[18]

One can imagine the auction as proceeding in four steps (see figure 12.1).[19] First, the bidder starts with the assumption that all valuations are the same.[20] This is the β_0 strategy component. Second, the price increases due to the auction mechanism, which causes bidders to exit once the price exceeds their own reservation prices. Third, the bidder observes the price at which competing bidders drop from the auction, and infers their types from that information by inverting their bidding functions (which, to repeat, all take the same form; therefore, all remaining participants are also performing the same recalibrations in precisely the same way). Fourth, the bidder amends his valuation (represented by the valuation function $\bar{v}$) by replacing each s with estimates of competitors' types, $\hat{t}^{(n)}$.

The bidder should continue to bid so long as the estimated value (the expected value of winning the auction conditional upon all other bidders dropping out) is greater than or equal to the price of the good. Viewed from the standpoint of the agents, the key operant criterion is "no regret": agents dropping out of the auction will have done so because price has exceeded their valuations; the agents who do not drop out of the auction will not regret remaining in. In equilibrium, all agents' strategies will have this no regret feature. The way this feature is incorporated into this model is that at every stage, bidders ask, "what happens if all the other bidders drop out, before I have time to react?" The bidders assess this possibility by assuming all bidders still active share

$$\beta_0(s) = \bar{v}(s,\ldots,s)$$

"No regret": At beginning, all valuations assumed the same

$$\beta_n(s,p_1,\ldots,p_n) = \bar{v}\left(s,\ldots,s,\hat{t}^{(N-n)},\ldots,\hat{t}^{(N-1)}\right),$$

Lowest price bidder of type s will drop when n bidders have already dropped   No regret criterion   Estimated types of all bidders who have previously dropped out

where $\hat{t}^{(N-k)}$ solves

$$p_k = \beta_{k-1}\left(\hat{t}^{(N-k)},p_1,\ldots,p_{k-1}\right)$$

k^{th} bidder's strategy when k-1 bidders have dropped out (from the standpoint of k, who just dropped)

$$= \bar{v}\left(\hat{t}^{(N-k)},\ldots,\hat{t}^{(N-k)},\hat{t}^{(N-(k-1))},\ldots,\hat{t}^{(N-1)}\right).$$

No regret criterion   Estimated types of all bidders who have previously dropped out

Figure 12.1. An Equilibrium Bidding Strategy, According to the Bayes-Nash Approach.

the same valuations.[21] We refer to this program as the Bayes-Nash School of Design because this program recasts the bidders' beliefs as Harsanyi-style "types," assumes that agents use Bayes's Rule to adjust their valuations to newly available information, and seeks to identify an equilibrium of consistent bidding strategies, given a particular auction structure.[22]

Circa 1970, those interested in this approach, such as Michael Rothkopf (then at Shell's Applied Mathematics Department) and Ed Capen (at Atlantic Richfield), would have understood its purpose as assisting their clients to formulate winning bidding strategies in existing auctions. Here, we can observe how the business school–private sector context inflected earlier Cowles information-processing enthusiasms. From the very beginning, the work of the Bayes-Nash School was motivated by appealing to a bidding irregularity that has come to be known as the "winner's curse." The curse purportedly results from the *ex ante* expected value of the commodity being greater than the expected value of the commodity upon learning that the bidder has won the auction. Winners are "cursed" because they win only by overestimating the value of the commodity.

Unsurprisingly, the classic example of such behavior was bidding for offshore oil tracts (the common value auction used to go by the name "mineral rights" auction), and the first "application" of the Bayes-Nash approach was to assist oil companies bidding on such tracts. [23] The classic description of the winner's curse had appeared in the *Journal of Petroleum Technology*. However, the relationship between this bidding expertise and game theorists' ambitions turned out to be awkward. Why would rational bidders need guidance in formulating their strategies?

Since that time, the Bayes-Nash School has also insisted it can leverage its expertise to go one step further and restructure markets. Here, the desire of governments to substitute regulation and bureaucracy with methods of market allocation superimposed a crucial consideration, as it has in so much recent market innovation. Early in the 1990s, Robert Wilson had gotten involved in the effort to devise markets for electricity; around that time, as we will later discuss, several members of this school involved themselves in efforts to auction spectrum licenses for the U.S. Federal Communications Commission. Shortly thereafter, enthusiasm for auctioning off public assets went global in scope.

Although the associated goal has often been couched in the language of seller revenue maximization, in fact this involved a shift in viewing markets as a technology for assisting agents in inferring the correct value of the commodity in question. Now, purpose-built markets could be enlisted to help agents think. "Truth" was now located both "out there" and "in here"—at least once the market had done its work. Truth was "out there," in that the market was designed in such a way that the price of an appropriately designed auction equaled the putative objective monetary value of an item for sale; it was "in here," in that bidders were sophisticated enough to infer this superior value from the "signals" conveyed by other bidders during the auction process, avoiding any behavior they would later regret, at least in equilibrium. As we have insisted in our discussion of the abstract information space, information at this juncture was being treated as, at least in part, inaccessible to the isolated individual knower. The Market was conceived as getting at things no individuals were aware they knew, at least at the outset of the process.

To what extent were these developments tracking developments in real markets? Following Vickrey, the Bayes-Nash School has traditionally focused on the properties of stylized representations of the "basic" auction forms, and has distinguished between markets in terms of a small handful of features. Therefore, one task of this school has been to taxonomize the relevant market formats with regard to their impacts on information. Take, for example, the textbook example of an English auction: one describes it as an open (versus closed) auction; if there is more than one item for sale, one should specify whether the auction is simultaneous (versus sequential), uniform price (versus discriminatory), and independent (versus combinatorial). One might also factor into the analysis whether the auction would require reserve prices and entry fees. Explicitly specifying each of these features would presumably help the analyst to nail down the properties of the market.

For example, the Bayes-Nash School has often praised the English auction for optimality properties stemming from its ability to facilitate bidder learning. Of course, this was a highly stylized representation of the auction. Orley Ashenfelter has pointed out that almost no auction operates according to the rules of the "English auction" actually studied in the literature, and certainly not real-life English auctions.[24] Economists responded to government efforts to structure markets not by changing their description of markets to track some external evolutionary trajectory but, instead, by attempting to build their own versions of markets for clients and make them a reality.

But these attempts raised a host of complicated issues. Was there really little else one could say about the process of facilitating learning in markets other than to claim that English auctions did it best? But if so, then why would one need economists' expertise? Another sticking point would be to explain why English auctions were relatively scarce in the larger market ecology if they were so superior. One response was to insist that private and common values were polar cases, with most commodities falling somewhere in between, which would necessitate that different market formats be tailored for different purposes and settings. However, ensuring that bidders make the required inferences becomes problematic for information settings other than the supposed polar "private" and "common" value cases.

A lucid summary of several negative results puts it this way: "Auctions in environments with multidimensional signals are often inefficient because of the impossibility of efficient aggregation of multidimensional information in a one-dimensional bid."[25] This seemed to not bode well for aspirations of a broad scope in the activities of these economists. The severe limitation in scope has forced the Bayes-Nash School to substitute piecemeal analysis for the traditionally accepted standard of mathematical proof making when making pronouncements about the optimality properties of many auction forms. Consequently, game theorists adopted the slogan: you couldn't just look up market design in a book. Real market design required (unspecified) experience. All that game theory you learned in grad school? Never mind!

Among market designers, there was a general awareness of this problem, but few drew out its full implications. Experimental economists were the exception to this generalization; as we will see in the next chapter, Vernon Smith took the Bayes-Nash program to task for failing to meet even its own internally generated standards. Unsurprisingly, members of the Bayes-Nash School have responded with a few salvos of their own:

> [Game theory] was a huge movement because it was a general method for treating all kinds of social institutions. But people don't always play Nash equilibrium; we are not sure why in some games they do and in other games they don't. And so this general method that has raised our sights and aspirations is highly imperfect and that represents a real challenge for how we should think about these things. Certainly, there is a division now. There are some experimentalists who think they have the answer: you simulate an institution in a laboratory. I don't agree with that. Laboratory simulations are, for the most part, based on lots of assumptions. For example, there are experiments going on using auctions, which involve complex auction institutions. But which details should be modeled in the lab, what should the values of the bidders be, and how should they be related to each other? It is easy for experiments to miss the

main point. . . . And so working out the role of experiments in economic theory is a big challenge.[26]

Milgrom's complaint is that due to the presence of untested assumptions necessarily appealed to in the experimental setup, one is at a loss to know how to move from experimental results to a general theory. But it would be hasty to assume that everyone would necessarily agree that arriving at a general theory in Milgrom's sense should be the "main point" of market design, as we will see in the next chapter.

[13]

THE EXPERIMENTALIST SCHOOL OF DESIGN

Let us first acknowledge the obvious: claiming that experimental economics constitutes a distinct school of market design is likely to strike some contemporaries as rather odd. After all, isn't experimentation about making economics more scientific by subjecting theoretical claims to controlled testing? But Experimentalists have harbored far more vaunting ambitions. One gets a sense of these ambitions by examining the backgrounds of Vernon Smith, Charles Plott, Stephen Rassenti, Robert Bulfin, and Alvin Roth.

The first thing to notice about members of this group is that they did not trace their genealogy out of some well-established social scientific experimental tradition, such as that found in psychology, but instead hailed from engineering and operations research. In light of this background, it begins to make sense that such figures would also occupy themselves with problems of economic design. And then there are additional reasons.

The *Experimentalist School of Design* has roots in the earlier work of the Walrasian mechanism designers, the development in engineering departments of optimization routines, the development within economic laboratories of computerized experimental methods, and the neoliberal field of "public choice." As we discussed in chapter 9, the first contribution of Experimentalists to attract the rapt attention of the orthodoxy (perhaps we should call it experimental economics' "killer app") was *not* the testing of economic theory but, instead, the development and deployment of novel market forms to displace bureaucratic decision making. It is only by keeping in mind these ambitions that (for example) the seemingly inexplicable decision to entitle a 1994 *Economic Theory* symposium on laboratory experimental methods "designer markets" begins to make sense. In the introduction to that symposium, Plott declared:

> Designer markets are becoming a reality. A merger of theory and experimental work is setting stages for a different kind of economics. The modern theory of mechanisms suggests that it is possible to design markets and/or decentralized mechanisms that can perform tasks that were thought to be impossible. The mechanisms themselves can become active participants with computers solving complex optimization or coordination problems based on "messages" submitted to the system by decentralized agents.[1]

What was this new species of economics, wherein the mechanism was conceived as an active participant? The most important early example was Steven Rassenti's (1982) dissertation at the Department of Systems and Industrial Engineering at the University of Arizona, which was taken under Vernon Smith and the engineer Robert Bulfin. The subject of the dissertation was ostensibly the development of an algorithm to solve a problem in optimization theory known as "0-1 integer programming," but in fact it professed an ambition to create specialized markets for cases where there are multiple indivisible commodities to be allocated, and where valuations for those commodities are complementary.

The commodities to be allocated in that instance were airport takeoff and landing slots. As discussed in chapter 9, the U.S. federal government had in 1978 deregulated airline routes, leading Charles Plott and his co-authors to develop a "market" method to replace the committees that had been formerly responsible for allocating such slots. Prior to this, in 1975, a graduate student named Arlington Williams had begun work on computer-assisted markets, for the purpose of improving the recordkeeping for classroom experiments held at the University of Arizona.[2] Later still, Vernon Smith hit upon the idea of using computers to accomplish the complicated allocation task studied by Plott, and proposed that Rassenti devote his dissertation to doing so. Smith recollects that, in the resulting dissertation, "Stephen [Rassenti] had created the concept of the smart computer assisted exchange."[3]

Once again, it is possible to gloss this innovation as primarily driven by the idiosyncratic nature of the commodity to be allocated (to the list, we now add multiple commodities, some with "complementarities"), and given that such problems were endemic to the Walrasian program, it is tempting to do so. But as in the case of the Bayes-Nash School, focusing on the peculiarities of the commodity to be allocated would cause us to miss the larger point. Under the new regime, the market, including its rules and participants, would now be explicitly conceived as a "person–machine system," a hybrid computational device.[4]

Experimentalists nowadays proclaim that everywhere, from the trading pit, to the regulator's office, to the corporate boardroom, can benefit from a little "economic system design."[5] In one respect, the first adjective is a bit of a misnomer, or at least imprecise, since for modern Experimentalists there is no delimited "economy" that serves to circumscribe their attentions; hence, for the Experimentalist School, unlike in the cases of our previous schools, there can be no canonical model of the economy. There is, however, a generic "set packing problem" that results in complications that must be addressed in order to successfully design an economic system.[6]

Here, we examine one canonical instance of this problem, interpreted as an auction of multiple items for which agents may have complementary valuations. The auction is populated by a set of agents N, who wish to acquire various elements of M, the set of objects for sale. Each agent $j \in N$ submits a bid for a subset S of M, $b^j(S)$; the highest of these bids on S is $b(S) = \max_{j \in N}$. If the highest bid on S is accepted by the procedure, then x_S is set equal to 1, and 0 otherwise. Then the problem can be formulated as follows:

$$\text{Maximize} \quad \sum_{S \subseteq M} b(S) x_S$$

$$\text{Subject to} \quad \sum_{i \in S} x_S \leq 1 \quad \forall i \in M$$

$$x_S = 0, 1 \quad \forall S \subseteq M$$

The constraints ensure that no single object gets assigned more than once. This approach to market design regards markets as combinatorial optimization problems, which means they assign prices to various subsets of commodities for the purpose of maximizing the objective function. The distinctive mathematical feature of this maximization problem is that because bids for packages are permitted, solving the maximization problem involves properly assigning prices to disjoint sets of items.

One feature of this approach is an intensified and more sophisticated focus on the algorithmic properties of the market than is offered by the other schools of design.[7] (It was Rassenti's expertise in algorithms that led Smith to suggest reorienting his dissertation to designing "smart markets."[8]) When stressing the computational properties of market operations, the Experimentalist School appeals to the "computational efficiency" of algorithms. Using the classification scheme of mathematicians, the sort of optimization problem studied by Rassenti—commonly referred to as a "set packing" problem or a "knapsack" problem—is NP-complete.

Though not necessarily intractable, NP-complete problems are regarded as too computationally burdensome (or computationally "inefficient") to solve directly, with the implication that one needs to employ a simpler (i.e., more "efficient") approximation algorithm to get the job "nearly" done. When focusing on the algorithmic properties of markets, this approach recommends substituting less computationally burdensome procedures, often by shifting part—though not very much—of the computational burden onto the "human persons" (bidders, in the case of markets) to assist in the search process.[9] For example, an "improvement bid mechanism" can be regarded as a relatively simple subroutine that organizes bids into "accept" and "reject" categories and then carries out an update on who qualifies as the highest bidder.[10] Markets, once conflated with the act of exchange, are now credited with being able to solve immensely complex maximization problems. The relative status of humans versus "mechanisms" in this process becomes inverted in the quest to overcome complexity.

For the Experimentalist School of Design, information was neither lodged in agents' heads (as it was for the Walrasian School) nor both "out there" and "in here" (as with the Bayes-Nash School), but instead abides as states in the computation of the market. Although they did talk about offloading some of the computational burden onto agents, members of the Experimentalist School did not view the "person" part of this "person–machine system" with much in the way of cognitive capacity.

Perhaps some of this attitude derived from the experience of manipulating students in experimental settings. The dreaded hive mind of collective consciousness had finally made its appearance in orthodox economics. Agents are shape-shifters in the Experimentalist School—sometimes viewed as incapable of coping with the substantial computational requirements imposed on them by Bayesian inference; in some other cases (like the Gode and Sunder work, discussed in chapter 8) they can't think at all. People may be smart, stupid, or anywhere in between in the New New Economics. But prudence dictates it is best to assume the worst and to ensure that the performance of markets is robust to the cognitive capacities of agents, or lack thereof. Such robustness is accomplished by offloading most of the task of information processing entirely onto the market mechanisms. The economist's task is now to build markets to handle the cognition that agents cannot—or, to use the highly appropriate term favored by experimentalists, to build "smart markets."

Whereas game theorists have historically limited themselves to considering the properties of a handful of stylized market forms, experimental economists have tended to explore the properties of the machine as a whole. This explains the penchant for referring not to "open" and "sealed bid" auctions, or "first-" and "second-price" auctions but, rather, to "adaptive user selection mechanisms" or "national resident matching programs." Markets are no longer described in terms of some imagined generic general properties; instead, they are viewed as specific devices built to specifications.[11] In the realm of markets-as-algorithms, anything ranging from the minutiae of correct coding to the construction of effective graphic user interface could, in principle, affect the attainment of goals.

This is not to say that the Experimentalist School has *no* theories of markets guiding it. Experimentalists often conceive themselves as sharing a distinct theory of markets, although this school has not in practice consistently grounded its approach in a single set of principles. Perhaps unsurprisingly, this school frequently refers to existing programs in orthodox economics when justifying its solutions to the profession. When wearing a Walrasian hat, it notes that complementarity produces a nonconvexity in the consumption set, which, if serious enough, rules out the existence of a competitive equilibrium and argues that competitive equilibrium prices no longer suffice to coordinate agents to optimal allocations.[12] A thoughtfully constructed user interface might be just the trick to accomplish such coordination, however.

When wearing a Marshallian hat, it relies on the maximization of "allocative efficiency" or "system surplus" or, mixing the two terms, "system efficiency." However, this goal serves less as a faithful representation of Marshall's thought than as a convenient rhetorical choice: "the concept of efficiency frees the experimenter from some of the constraints placed on progress because of the lack of an adequate theory."[13] Yet when considering computational features, the headline goal frequently becomes that of reducing computational complexity—something nowhere to be found in conventional microeconomics textbooks. The latter concession is especially noteworthy, because it is grounded in a distinct approach to information, untapped by either of the other two schools.

At times, practitioners would seem to suggest that they had discovered a Philosopher's Stone to synthesize all the various economic approaches to markets in such a way as to address computational concerns. Alvin Roth is the best example of an experimentalist who has attempted to define market design as a universal nostrum. In his telling, market design is rooted in the subset of game theory known as "matching theory," and in the work of Gale and Shapley on "stability" in particular.[14] But it would be a mistake to credit this group with forging a grand synthesis between Walras,

Marshall, game theory, and machine theory; the truth is, their practice doesn't conform to any single, underlying theory.

More typical of the Experimentalist School is to take strong exception to the claims advanced by their primary rivals in the enterprise of market design:

> In a path-breaking theoretical contribution, Milgrom and Weber (1982) introduced the concept of the common value auction, as distinct from the independent private values auction. But the path broken, leading to important theoretical insights, turned out to be littered with potholes in practical applications not sensitive to the negative aspect of design criteria that focused only on the common value issue.... Milgrom and Weber (1982) assume that "To a first approximation the values of these mineral rights to the various bidders can be regarded as equal, but bidders may have differing estimates of the common value." Such approximations are critical to the development of theorems, but anathema to design applications.... A bidding equilibrium does not exist in auctions defined in [a] mixed environment. The difficulties are foundational.... An important implication of this is that it cannot be claimed that, although we have no equilibrium theory . . . we can still apply game theoretic "intuition," developed from complete information examples, to articulate useful guidelines about market design.[15]

Here, Smith attacks the paper introducing the canonical model of the Bayes-Nash School, and voices the very same objection we previously encountered in chapter 12—namely, that there are no equilibrium-theoretical results for environments where values consist of private and common elements—which is to say, for every single environment that would interest a market designer.

Game theorists do sometimes grudgingly admit this, but Smith also rejects their common immunizing stratagem, that even where game theoretic methods generate no usable results, such methods can still provide useful intuition.

Experimentalists often suggest that because individual behavior is not invariant to the institutional and environmental setting, the search for any single, general model of individual choice (such as those offered by game theorists) is likely to be frustrated: "The nature of the design process dictates that process performance at the aggregate level attracts the most attention. Activity at the individual level of analysis is frequently more complicated and less consistent with [economists'] expectations."[16]

One of the most noteworthy consequences of focusing on machine performance and distrusting the computational abilities of the human person is the reduction in the importance of nailing down any specific model of the agent. While some experimentalists do sometimes attempt to test for the accuracy of models of individual choice, what stands out to the reader of this literature is the easy coexistence of many such cognitive models in their rationalizations; sometimes agents are portrayed as if they form their expectations adaptively, sometimes in an "unbiased" fashion, and sometimes consistent with "rational expectations." More fundamentally, experimentalists object to the treatment of *information* at the hands of game theorists: "The information and computational requirements of game theory offer little insight toward engineering rules of trade or in guiding strategic behavior of intelligent agents."[17]

Even Alvin Roth has acknowledged that upon being asked to design the medical resident matching program for which he is best known: "none of the available theory ... could be directly applied to the complex market.... The only theoretical parts of [my] book [on matching] that applied directly to the medical market were the counterexamples"[18] Among other drawbacks, (game) theory was unequipped to deal with computational issues. Experimentalists would act as *bricoleurs*, creating their markets out of a variety of materials, machines, and men.[19] Whereas the Bayes-Nash School engages in *taxonomy*, as a prelude to *studying* the properties of various markets, the Experimentalist School uses a *toolbox*, as a prelude to *building* markets.

[14]

HAYEK AND THE SCHOOLS OF DESIGN

In the previous chapters, we have shown how changes in markets and in notions of information have driven the concerns of the profession. We began by suggesting that Hayekian arguments about the significance of agent epistemological capacities were equally important for all involved, but while covering the three Schools of Design in chapters 11, 12, and 13, we omitted making that case. Now is the time to usher Hayek back into the inner sanctum of our narrative.

It is perfectly licit and normal by the twenty-first century for an economist in good standing to acknowledge the importance of Hayek. Take, for example, the 2007 Bank of Sweden Prize winner, Eric Maskin:[1]

> Hayek had a remarkable intuitive understanding of some major propositions in mechanism design—and the assumptions they rest on—long before their precise formulation. Indeed, his understanding seems to have been a guiding influence in their formulation.[2]

Maskin's two "Hayekian" propositions are, first, that "competitive markets are informationally efficient" and, second, that "the market mechanism is uniquely incentive compatible." Notwithstanding Hayek's intuition, a firm grasp of formal economic analysis (particularly game theory) eluded him, preventing him from grasping the nettle ("he did not anticipate—as far as I can tell—the Vickrey-Clarke-Groves mechanism for determining a Pareto optimal public goods allocation"). Nevertheless, Maskin describes him as a "precursor" and a "guiding influence" (as if these would serve as the same thing), even going so far as to make the interesting suggestion that those most involved in developing the game theoretic literature on markets did so with Hayek in mind.

Unfortunately, Maskin never seriously pursued this idea any further in this work—in all probability because he exhibits no more than a basic, bare-bones understanding of the corpus of Hayek's work. Unsurprisingly, both passages Maskin cites in support of his interpretation of Hayek were taken from the same article, "The Use of Knowledge in Society"; neither says anything about "incentive compatibility," nor does Maskin feel impelled to provide a single, specific example of Hayek's guidance.

While the orthodoxy's lack of curiosity concerning its history is in no way surprising, one might have hoped for better when it came to the self-appointed caretakers of Hayek's legacy: the Austrians. But to date, their efforts to address Hayek's influence on the orthodoxy have proved no more insightful. This was nowhere more apparent than in the aftermath of awarding the 2007 Bank of Sweden Prize to Leonid Hurwicz, Eric Maskin, and Roger Myerson. Initially, some Austrians greeted the occasion with applause, as an acknowledgment of Hayek's worth so incontrovertible as to be undeniable by even the most blinkered orthodox economist.[3] But this position apparently ran up against the perceived need to maintain the distinctiveness and independent virtue of the

Austrian approach, not to mention the traditional insistence upon the "articulate" versus "inarticulate" knowledge distinction (which was often used precisely to upbraid Walrasians such as Hurwicz).[4] So, subsequently, some Austrians executed an about-face and now accused both Walrasians *and* Bayes-Nash game theorists of "failed appropriation" of Hayek.[5]

One might expect that this turnabout would stimulate an interest in pinpointing exactly what it was that game theorists sought to appropriate and why.[6] To that end, some Austrians did organize a conference at George Mason University, with the laudable intention "to examine and provide us with insights into the impact of Hayek's work on the research direction of other scholars in economics and political economy ... [to] stimulate a conversation about the deep impact of Hayek's ideas."[7] But so far this project was hampered by an apparent commitment to a single, monolithic "Hayekian framework" about which mainstream approaches to the "economics of information"—putatively characterized by a flawed adherence to the "omniscience" of economic agents—could be said to have misunderstood. If instead the real Hayek had changed his mind (as we argued in chapter 6), as had the economic orthodoxy, then something would inevitably get lost in the dance of influence. Unfortunately, the ahistoricity of their privileged approach has induced the Austrians to miss the most direct avenues of Hayek's "deep influence" on orthodox economics.

We have already noted the links between studies of information in markets, the Socialist Calculation Controversy, and Hayek—most directly in the career of Leonid Hurwicz. It is no exaggeration to point out that Hayek motivated Hurwicz in a fundamental manner to initiate the Walrasian School of Design. Hurwicz has acknowledged Hayek's influence, and cited his work often throughout his career. But that was no one-off occurrence; it so happens that members of *every single* school of design have explicitly and repeatedly motivated their work by referring to Hayek and the Socialist Calculation Controversy.

Staying with the Walrasian School for the moment, in the work of Kenneth Mount and Stanley Reiter, Hayek's concerns were expressed as follows:

> As Hayek saw the problem, economic information is naturally initially dispersed among economic agents ... and, in order to arrive at (optimally) coordinated actions, this information must somehow be communicated among agents.... Some methods for achieving that optimal coordination were regarded by [Hayek] as infeasible—e.g., transfer of all relevant data to a central planning board.... While his detailed discussion dealt almost exclusively with the competitive model as against one of a centrally planned economy, Hayek recognized the possibility of rational design of the institutional framework and the possibility of new economic institutions ("new" in the broader sense of "hitherto not conceived" as well as "other than those historically observed").[8]

The information that was required to coordinate economic action was dispersed, necessitating that this information somehow be communicated. For Hayek, such communication outside the price system was infeasible, but according to Mount and Reiter's reading, it was not *impossible*, and perhaps was even feasible for some institutional framework not akin to central planning. Prices could convey information, as Hayek had contended, but there was no reason to restrict such economic communication to prices—and, in light of the problems created by nonconvexities and indivisibilities, there would seem to be excellent reasons to consider alternatives. Markets were supposed to give people what they wanted; but this would not happen with certain kinds of commodities and certain kinds of preferences (and cost structures) attached to them.

The informational claims of Hayek were also foremost in the considerations of members of the Bayes-Nash School of Design. We noted earlier that Vickrey wished to critically examine Abba Lerner's counterspectulation method in the context of central planning. But Vickrey was far from the only member of the Bayes-Nash School of Design to acknowledge the influence of Hayek. Consider Robert Wilson:

> A half-century ago, Friedrich von Hayek offered a new perspective on markets, prices and the invisible hand. In his view, the fundamental process of a market economy is price formation. He interprets prices resulted from competing bids and offers as summaries of information dispersed among traders.... A quarter century later, the developers of the Economics of Information discovered that market imperfections attributable to informational asymmetries can cause serious inefficiencies.... Initially, the main tool was price theory, but more recently it has been game theory. In particular, it is the flavor of game theory that originates in the work of ... John Harsanyi.[9]

And, in a survey article on the Bayes-Nash School, consider the more specific reference linking a result of the school to the work of Hayek:

> It is often pointed out (for example, by Hayek ...) that one of the remarkable and important features of the price system is its ability to convey information efficiently. All that a buyer or a

seller needs to know about a commodity's supply or demand is summarized by a single number, its price. Does the process of price formation by competitive bidding have such information efficiencies? In the common-value model, the bidders lack complete information about the item's true value; each bidder has different partial information. However, even though no single bidder has perfect information, it can be shown that, if there is perfect competition in the bidding, the selling price reflects all of the bidders' private information.... Thus the selling price conveys information about the item's true value. With perfect competition, the price is equal to the true value even though no individual in the economy knows what this true value is and no communication among the bidders takes place.[10]

Note well that these leading members of the Bayes-Nash School made reference to the work of Hayek in the context of interpreting the overall significance of their own achievements. Agents' knowledge was portrayed as difficult to access by the auctioneer/central planner, but it was also difficult for the agents themselves to access it; by contrast to the conventional Walrasian view, agents' knowledge was *untrustworthy*. But while agents' knowledge was deemed untrustworthy, they were competent enough to still be able to incorporate more information into their valuations, and therefore were deemed capable of highly sophisticated reasoning. The market would provide the information needed to carry out such sophisticated reasoning.

Experimentalists also framed their interventions by referring to the ideas of Hayek. One observes such framing not only in the "Hayek hypothesis" of Vernon Smith, discussed in chapter 8, but also in experimentalists' activities in making smart markets:

> The objective is to combine the information advantages of decentralized ownership with the coordination advantages

of central processing.... In effect we offer a solution to the Lange-Lerner-Hayek controversy of the 1930s."[11]

In discussing "information advantages," the importance of "central processing" is paramount. Central processing enables the completion of trades too complicated for individuals to complete on their own: "There is a puzzle as to the processes whereby our brains have [market exchange] and other skills so deeply hidden from our calculating self-aware minds."[12] At the hands of the Experimentalist School, the market is redescribed as a "price discovery" process, in almost direct parallel to Hayek's late discussion of competition as a discovery process. The change in language reflects the Experimentalist view that only skillfully designed markets—smart markets—can find the economic knowledge that cognitively limited agents are *incapable* of knowing.

This view is highly compatible with the notion that markets have the power to know things that agents cannot, the position taken by Hayek during his third period:

> In a 1968 lecture, "Competition as a Discovery Procedure," Hayek says "I propose to consider competition as a procedure for the discovery of such facts as ... (otherwise) would not be known to anyone...." Great insight; experiments have long demonstrated Hayek's proposition. People discover a price that they didn't know existed.[13]

And in direct reference to the ability of skillfully defined markets to substitute for human cognition:

> Human interactive experiments governed by a computer network enabled the accommodation of far larger message spaces, opened the way to the application of coordination and

optimization algorithms to the messages of subjects, and facilitated their capacity to reach sophisticated equilibrium outcomes they did not need to understand.[14]

Economic designers had managed to convince themselves that they had faithfully come to grips with Hayek's concerns.[15] The Walrasian School viewed itself as assisting the government in a number of areas, proposing rules for planning and suggesting the information to gather. Agents knew their "private" information, but the government did not. This established for the Walrasian School the task of "rational design of the institutional framework," which would amount to a novel kind of economic-cum-communication system—not central planning, but not quite like the market, either.

Initially, economists conceived themselves as designing various methods to help gather information; knowledge was held by dispersed agents, and the job of the economic designer was to figure out how best to transport knowledge from where it was to where it wasn't—lest the economy not operate properly. Designers of the Bayes-Nash School tasked themselves with helping agents come to know *their own* values. This they would do by helping agents to correctly infer values, and then by recommending the use of knowledge-enhancing auction forms. In an environment where economists increasingly found themselves selling their expertise, such ambitions carried considerable appeal, as we will see in the next chapter. Although game theorists attributed immensely impressive prodigious rationality to their version of the agent, nevertheless economists managed to carve out a special role for their own activities.

Finally, the Experimentalist School viewed the task of designers as constructing machines to discover an elusive knowledge that individuals could not discover or otherwise comprehend themselves. These smart markets would include people, but in practice would substitute for the judgment of people—for example, by replacing regulatory bureaucracies. The Ghost of Hayek haunted them all.

There were a few specific structural conditions in the history of the American economics profession that would have brought Hayek's concerns to the forefront and kept them there. We have touched on them in previous chapters. Some involved restructuring at the departmental or university level, such as the reinvigoration of Purdue's economics department and (later) the reinvention of Arizona's economics department and Caltech's program in social sciences as centers for experimental economics; others involved the establishment of new programs, such as Northwestern University's Managerial Economics and Decision Sciences (MEDS) department and its Center for Mathematical Studies in Economics and Management Science (CMS-EMS). But efforts spanning different universities, forging communities of scholars around topics related to information and economic design, would also constitute an important part of the story. In some cases, like the establishment of the Economic Science Association, one specific approach to market design was lent prodigious support.[16]

But, then, there were the yearly National Science Foundation Conferences on Econometrics and Mathematical Economics (NSF/CEME) Decentralization Conference Series, which would explore all the various permutations of information and design, giving us a glimpse into how Hayek's provocations came to be reinterpreted over time by members of each of our "schools." According to its former coordinator, "the mechanism design literature . . . stemmed from the [meetings of the] decentralization series."[17] A memo summarizing the proceedings of its first meeting (in 1971) noted that it covered:

(1) studies of specific "adjustment mechanisms" designed for particular problems and/or environments, and exhibiting "decentralization" in some sense or other;
(2) the development of formal definitions of decentralization, or of various aspects of decentralization, and attempts to define the circumstances under which one could say that one system is "more decentralized" than another; and
(3) the optimal use of information in decision-making systems, taking account of the costs of information and/or the given capacities that the decision-making system has for information processing.[18]

The initial proceedings reflected the concerns of the Walrasian School of Design, whose members dominated its roster of participants. It was first coordinated by Roy Radner in 1971 (who was a member of the Cowles Commission), and over the years tended to include a relatively small, though expanding, roster of repeat participants.

During the first ten years, Leonid Hurwicz topped the list, with eight papers delivered over that span, followed by Roy Radner (6), John Ledyard (5), Theodore Groves (5), and Thomas Marschak (4). By the time of the first meeting, Hurwicz had already begun to circulate his first early results on the nascent Bayes-Nash approach. At that meeting, Radner invited Hurwicz to share this work; subsequently, the series became an important forum for the "designer" community to develop and circulate work addressing Bayes-Nash themes.[19]

By 1980s, as the roster expanded to include such figures as Robert Wilson, Paul Milgrom, and Eric Maskin, studies explicitly cast in terms of game theory assumed increasing prominence. Explorations of computing and complexity made their appearance soon thereafter, addressed first by Mount and Reiter in the early 1980s, followed by other related scholars with increasing regularity.[20]

While explicit awareness of the nature of the original Hayekian provocation definitely diminished as the circles of scholars widened, it never entirely went away—possibly due to the more or less continuous presence of Hurwicz, John Ledyard, Thomas Marschak, Roy Radner, Stanley Reiter, and often other original, inaugural participants.[21]

Although one can find at those meetings representatives of each of the various schools still with us today, the economics profession has tended over time to move along the diagonal ray toward the "northeast" of the Information Space we introduced in chapter 10 (included here as figure 14.1), toward the region occupied by the Experimentalist School of Design.

Numerous considerations account for the profession's trajectory in this space. Prominent among them was the lack of success that members of the Walrasian and Bayes-Nash Schools had in fully reconciling the imperatives growing out of their own programs. One thinks immediately of Mount and Reiter, who got their start in the Walrasian School, but in their desire to address computational concerns were brought to reject central tenets of this approach. For instance, there was Alvin Roth, who had studied under Robert Wilson, but soon thereafter found his work on game theory to be of limited use, and so eventually adopted the methods of the Experimentalist School. To the extent that members of the Walrasian and Bayes-Nash schools felt compelled to adopt

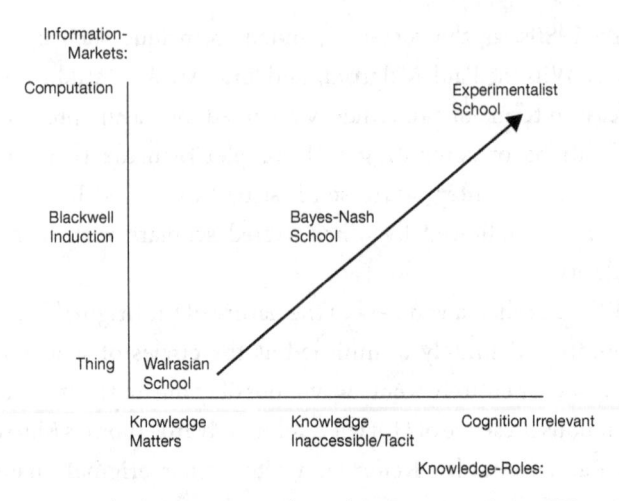

Figure 14.1. Orthodox Trajectory Through Information Space, III

computational themes, they often found they had to dispense with one or more central tenets of their respective approaches. And then there is the poorly acknowledged fact that economists have increasingly downgraded the significance of agent cognition for the operation of the market over time. Consequently, markets are less and less conceived as being about giving people what they may think they want, and increasingly about operating regardless of their wants, for the benefit of some entrepreneurial entity.

At this juncture in our narrative, we wish to signal a key departure from all those economics textbooks that want to claim it was the tool of game theory that was most responsible for the development of market design. Although a superficial reading of our previous narrative might seem to lend support for this position, we insist that viewing the history in this way fails to do justice to the historical record and obscures the trajectory of the profession. Over the course of the 1980s and early 1990s, Experimentalists had consulted for several government agencies (the Federal Aviation Administration, Civil Aeronautics Board, Federal Energy Regulatory Commission, NASA, the Arizona Corporation Commission); published numerous papers on market design for natural gas pipelines, electric power transmission, space station utilization, and (as discussed in chapters 9 and 13) airport landing slots; built market prototypes; and had even established a private company (Cybernomics) that would handle consulting contracts.

In 1991, Experimentalists' work on market design gained widespread attention, as the respected natural science journal *Science* published Kevin McCabe, Stephen Rassenti, and Vernon Smith's "Smart Computer-Assisted Markets."[22] And in what must have seemed like a formal acknowledgment of the increasing prominence of the Experimentalist approach, the Arizona Economics Department hosted the 1992 meeting of the NSF/CEME Decentralization Conference Series. That meeting included noticeably more experimental explorations than previous meetings. Of course, it also included plenty of papers located squarely within the Bayes-Nash tradition; yet, those paying close attention (including, presumably, the Experimentalists) may have discerned that a dark cloud for that program lay on the horizon: Eric Maskin delivered a paper containing the first in what would turn out to be a series of negative results for auctions with "multidimensional signals."[23]

While it would be an exaggeration to view these developments as flagrant evidence of a crisis of confidence in the Bayes-Nash approach, it is certainly no exaggeration to say that in the early 1990s, the Experimentalists would have been at least as likely as the Bayes-Nash adherents to be regarded as representing the vanguard in market design—a conjecture that may be confirmed by recalling that it was the specific efforts of Experimentalists that the landmark "Markets and Organization" report chose to single out for its praise.

Yet, as those with even a passing knowledge of the economics profession's recent history are surely aware, it would be the Bayes-Nash approach that within a few short years would come to be viewed by the press, students, and other outsiders as the highest expression of the economics profession's design ambitions. To understand this curious turn of history, it will be necessary to closely examine the single most important event that cemented market design as central to the identity of the economics profession: the U.S. Federal Communications Commission (FCC) spectrum auctions.

[15]

DESIGNS ON THE MARKET

The FCC Spectrum Auctions

Over the past three decades, market designers have argued that it is possible to reengineer markets to deliver any number of salutary results. Markets can reverse global warming, improve access to health care, redress racial and gender discrimination, or even accomplish "free lunch redistribution," while at the same time promoting allocative efficiency—so long as they are built correctly, in the opinion of some market engineers. It is a development that has followed policy, at least to some extent:

> Governments around the world have begun using markets as means to policy ends. Pollution control has been assigned to a market in emissions allowances. The right to use the electromagnetic spectrum for telecommunications has been auctioned off. In electricity supply, markets have replaced allocation by state agencies or regulated monopolies. In fisheries management, tradable quotas have started to be used instead of direct regulation.... Governments can successfully use markets. Information is the key. The market process—where it works well—generates information on which of the firms

are able to put scarce resources to the best use and on what the highest value use is. This information is unlikely to be revealed via a political or administrative procedure.[1]

While stressing the ability of markets to "reveal" information is a hallmark of the Bayes-Nash School of Design, Experimentalists also emphasize the ability to use purpose-built markets to promote policy. This justification for the role of economists has proved to be a hit not only within the economics profession but also across the social sciences, even with groups normally thought to be antagonistic to orthodox microeconomics.[2] The overwhelming impression is that is that market design has been wildly successful.

In the previous chapters, we provided a historical narrative of the intellectual development of this branch of economics. In this chapter, we bring into focus the circumstances surrounding the emergence of this new identity for economists to act as market designers. The single most important episode in the emergence of this new self-conception of economics was the U.S. Federal Communications Commission auctions for electromagnetic spectrum licenses. It is to this episode that we now turn.

THE CORPORATE RATIFICATION OF THE BAYES-NASH APPROACH

The successes of game theory in developing the FCC spectrum auctions have been trumpeted far and wide by governments, the media, and even Hollywood (in the film *A Beautiful Mind*). And they have been repeated within all tiers of the economics discipline, from the dense valleys of popular undergraduate texts to the rarified air of its journals. In the immediate aftermath of these auctions, almost every available account poured praise upon the efforts of the game theorists who had participated as consultants in the auctions, and marveled at the unprecedented involvement of academic economists in forging public policy. In economics departments, you hear about the FCC auctions every time a person delivers a paper on game theory or every time an economist lobbies for government funding during these austere times.

Those who have found their interests piqued by this application of game theory have been able to consult a handful of widely cited accounts. That these accounts were written by the academic game theorists who participated in the spectrum auctions ends up being pretty important. It is possible to summarize their accounts in the form of three lessons:

1. The decision to auction off electromagnetic spectrum licenses led to scientific considerations acquiring significance alongside political considerations.
2. The role of market design in (1) demonstrates its practical relevance for public policy.
3. The most compelling evidence of (2) can be found in the high revenues produced by the auctions, though market designers can reengineer markets to bring about any number of salutary outcomes.

The story began in 1994, when the U.S. Federal Communications Commission (FCC) commenced, for the first time, the practice of auctioning off spectrum licenses to the highest bidder.[3] The process of determining the best method of selling the rights to control certain frequencies of the electromagnetic spectrum was marked by another innovation: the heavy involvement of academic game theorists—practitioners of one of the most abstract mathematical fields of economics, often thought to exist at a remove from practical problems. Once the first set of auctions was completed, and the dollar tally came in, those economists gleefully took credit for what was initially perceived as a highly successful performance.

It is commonplace for the firsthand accounts of the FCC auctions to begin with a discussion of the stipulation of several goals for the auctions by the U.S. Congress. In fact, Congress charged the FCC with:

I. The development and rapid deployment of new technologies, products, and services for the benefit of the public, including those residing in rural areas, without administrative or judicial delays;
II. Promoting economic opportunity and competition and ensuring that new and innovative technologies are readily accessible to the American people by avoiding excessive concentration of licenses and by disseminating licenses among a wide variety of applicants, including small businesses, rural telephone companies, and businesses owned by members of minority groups and women;
III. Recovery for the public of a portion of the value of the public spectrum made available for commercial use and avoidance of unjust enrichment through the methods employed to award uses of that resource; and
IV. Efficient and intensive use of the electromagnetic spectrum.

The list represents the outcome of a prolonged debate over the role of government in promoting access, innovation, competition, and "competitiveness." The FCC, however, would eventually take the position that all these complicated considerations involving industrial organization, macroeconomics, and distributional equity should ultimately be reduced to the narrower goal of "economic efficiency," and that the most appropriate way to pursue this goal should be to award licenses to their highest valued users.[4] By replacing the goals of Congress with their preferred "efficiency" criterion, the FCC staff economists[5] were able to ground their policy analysis in game theory.

The true significance of this was not, as has been commonly asserted, the substitution of political with "scientific" considerations but, rather, the effective enrollment of a specific group of academic game theorists into the FCC's policymaking process. The appearance in the FCC docket of a call for game theoretic analysis of how best to award licenses to the highest valued user was unprecedented, and it gave certain interested parties the idea of hiring academic game theorists to further their objectives.

But those hoping to ground controversial public policy in uncontentious science would be disappointed, as the enlistment of an increasing number of economists to the market design process would result in a remarkably diverse array of inconsistent proposals—and ultimately, a failure to produce any clear-cut recommendation.[6] The sticking point was that game theory supplied no global discipline with regard to the type of recommendations tendered: a game theorist could legitimately support any of an array of auction forms by stressing one set of information properties over others. Participating game theorists did tend to conceptualize an auction as a Bayesian learning game, in the fashion of the Bayes-Nash School of Design, discussed in chapter 12. They focused attention on the release of information during the auction that would better promote knowledge of the licenses' true value, hence promoting efficiency.

There was, however, no conventionally accepted standard for determining the precise value of the information provided by a given auction, much less the "true" value of any good. This was a problem for attempts to generalize existing results to an environment with multiple heterogeneous goods, or in the argot of game theorists, multiple-good environments with "multidimensional types."[7] Game theorists therefore supported their recommendations not with their own conventionally accepted standards of mathematical proof, but with loose analogy and piecemeal analysis, mooted in seemingly clear but frequently contradictory catch phrases as "the more open, the better," or "make sure participants get quality information," or "avoid free-rider problems."

Participants in the run-up to the spectrum auctions have acknowledged that game theory was unable to provide a knockdown argument for the optimality of a specific auction form.[8] They attributed the lack of a determinate recommendation to a local disagreement over the magnitude of various effects, but this does not begin to get to the heart of the matter. The lack of a determinate recommendation was less a disagreement over the significance of various learning effects than it was a disagreement over the *aims* of the auction.

One source of disagreement was a methodological clash that accounts of the auctions failed to note, but that we are now in a position to appreciate: some market designers were members of the Bayes-Nash School; others were in the Experimentalist School. In this case, both game theorists and experimentalists were concerned with the presence of interdependent values of different geographic spectrum allocations, but they understood the problem interdependency posed in a radically different way. Experimentalists argued that the only sort of market algorithm that could be counted on to produce a dependably "optimal" allocation of licenses required a method for collecting information on the value of *packages*—or combinations of licenses—in addition to the value of individual licenses. Experimentalists judged this allocation problem to be formally equivalent to a "knapsack" problem in combinatorial optimization; called for package bidding; proposed a smart market to cope with the resulting processing complexity; and supported this recommendation with laboratory evidence for its (*ex ante*) optimality.[9]

By contrast, the game theorists who opposed package bidding[10] argued that merely asking for information on package values would ultimately *reduce* the amount of information collected. Package bidding would remove the incentive to bid on single licenses, reducing information on license values, suppressing the prices of individual licenses, and ending ultimately in a failure to displace a high package bid.[11]

To summarize, Experimentalists argued for a combinatorial auction and proposed a "person-machine system" to perform the complex processing task. Game theorists concentrated on bidders' incentives to release private information within a highly stylized Bayes-Nash auction game, and proposed independent license bidding.

But while there was ample room for disagreements over the efficiency properties of the auction proposals, or even the appropriate methodology for auction theory, the companies' narrowly constituted interests clearly played a major role:

> [T]he business world was fully aware of [the strategic significance of] the rulemaking process and had engaged many groups of consultants to help them position themselves. Businesses understood that the rules and form of the auction could influence who acquired what and how much was paid. The rules of the auction could be used to provide advantages to themselves or to their competitors. Thus a mixture of self-interest and fear motivated many different and competing

architectures for the auctions as different businesses promoted different rules.[12]

The most prominent "consultants" used by businesses to "position themselves" were game theorists hired by the large telecoms. Several companies responded to the FCC's rulemaking process by lobbying for preferred sets of auction rules, and some—mostly the monopolistic regional wireline telephone service providers created from the 1982 breakup of AT&T—enlisted economists to draft supporting comments. The telecoms went on a hiring spree: Nynex hired Robert Harris and Michael Katz of UC Berkeley; Telephone and Data Systems (TDS) hired Robert Weber of Northwestern; Bell Atlantic hired the Yale economist Barry Nalebuff and Jeremy Bulow of Stanford; Airtouch hired R. Preston McAfee from the University of Texas; Pacific Bell hired Paul Milgrom and Robert Wilson from Stanford.

It is clear that many aspects of the proposals funded by the telecoms were determined by their narrow acquisitive strategies. Consider the case of "package bidding":

> In the U.S. telecommunications spectrum auctions, sophisticated bidders anticipated the effects of packaging on the auction and lobbied the spectrum regulator [the FCC] for packages that served their individual interests. For example, the long distance company MCI lobbied for a *nationwide* license which, it claimed, would enable cell phone companies to offer seamless coverage across the entire country . . . *regional* telephone companies such as Pacific Bell lobbied for licenses covering areas that fit well with their own business plans but poorly with the plans of MCI.[13]

The authors should know: *one of them was the lobbyist for Pac Bell*. There are many other examples of "lobbying the spectrum regulator," including how to sequence the auctions, how to set collusion rules, and so on.

So, for both these reasons, market designers failed to produce a single recommendation. In an ironic twist, the task of determining the public version of what game theory had ultimately dictated fell to the FCC. The multiplicity of aims and proposals forced the FCC to display some creativity in conjuring a "consensus" recommendation for the auction form—the simultaneous multiple-round independent auction [SMRI]—given that it was the one that most economists opposed.[14]

Working out the details of the never-before implemented SMRI turned out to require far more elaborate competencies and redoubled efforts beyond those deployed in the initial rounds of the public policymaking process. (How are you going to rivet "simultaneity" and multiple rounds together to form a working machine?) Consequently, Experimental economists were recruited to participate in the design of the auction. They succeeded where the game theorists failed, in building operational auctions. This success was due in large part to their previous experience in constructing computerized markets, inside and outside the laboratory. *These* were the real machine builders.

Yet the game theorists took the credit. And for what? Game theorists have been loudly trumpeting their success in "designing [the FCC auctions] for multiple goals" for two decades, leading directly to the explosion of market design. And their claims have gone more or less unchallenged despite considerable evidence built up in the interim to the contrary.

It is demonstrably false that the spectrum auctions satisfied the congressional goals.[15] Many businesses buying licenses defaulted on their down payments,[16] leading to considerable "administrative delay" in re-awarding licenses.[17] The lion's share of licenses won by "small" and "entrepreneurial" businesses went to entities bankrolled by large telecoms, representing a failure to get licenses into the hands of a "wide variety of applicants."[18] The auctions did not live up to their promise to promote "rapid deployment [in] rural areas," as both large telecoms and smaller companies tended to concentrate their effort on large metropolitan areas.[19] Overall, the allocation of licenses produced by the auctions proved to be unstable, as the industry has gone through a spate of mergers, acquisitions, and bankruptcies, ultimately leading to a high degree of license concentration.[20] Commenting on some of these events, one anonymous FCC official candidly observed, "this certainly does make us look like a bunch of idiots."[21]

True, the auctions did capture a tidy sum for the government coffers—more, anyhow, than administrative hearings would have—but perhaps they did so at the expense of any solid foundations for the economic health of the industry. Furthermore, this focus on the billions of dollars in bids draws attention away from the role the consultants played in *decreasing* auction revenues. The consultants' efforts achieved their most spectacular result in the decision of MCI (a deep-pocketed nationwide bidder) to drop out of the December 1994 auction as a result of successful persuasion by economists of the FCC to reject nationwide package bidding.[22] Consulting economists argued against a "smart market" mechanism, which in experiments had produced higher revenues.[23] They advised their clients on forming consortia that had the effect of reducing competition in the auctions.[24]

On occasion, the consultants became involved in the auction process very directly. Immediately prior to the December 1994 auction, Paul Milgrom appeared on *CNN Business Morning* and proclaimed, "Pacific [Bell] expects to win licenses in California. We expect the other bidders to have an opportunity to become discouraged when they see how determined we are."[25] After the auctions had concluded, Milgrom reported that his appearance did "successfully discourag[e] most potential competitors from even trying to bid."[26] Rather than assigning credit to economists for raising large revenues for the treasury, it is probably closer to the truth to credit economists for helping their clients to acquire licenses at bargain prices.[27]

Hence, the true lessons of the FCC auctions are:

1. The decision to auction off electromagnetic spectrum licenses led to the commercial interests of a handful of large telecoms acquiring significance alongside both political and scientific considerations.
2. The role of market design in (1) demonstrates its ability to deliver clients a valuable service.
3. The most compelling evidence of the success of (2) is the market designers' delivery of licenses to their clients, relatively cheaply.

A primary justification for market design is that it would replace corrupt regulatory practices with transparent market-like procedures:

> Auctions, unlike administrative hearings, are transparent.[28]

> [A] well designed auction is the method most likely to allocate resources to those who can use them most valuably ... [and avoid] political and legal controversy, and the perception, if not the reality, of favoritism and corruption.[29]

Arguments that stress the transparency of markets relative to bureaucracy have a long history; interestingly enough, the FCC case has taken center stage in the history of economics: in 1959,

Ronald Coase argued for replacing FCC bureaucracy with markets. But market designers' arguments differed from Coase's in one important respect. It would not be enough to merely rejigger the definition of property rights and entrust their allocation to The Market: "For markets no less than for buildings, it is the details of design that determine whether or not they work well. Both God and the Devil are in the details."[30] Market design espouses the position, "markets work—but we can make them work better." It is not merely markets that are tasked by market designers to replace politics, but some combination of markets and *economics*.

One is counseled to be very suspicious of government bureaucracy on the grounds that it is prone to corruption. Markets are better, but it is usually best not to leave such markets to be generated spontaneously: "the Devil is in the details." Instead, one must employ principles of sound market design to ensure that goals are pursued effectively and with integrity. And since it is really only by experience that one grasps these principles, the best way to make sure sound design principles are employed is to enlist the help of a market designer. "God is in the details," as well. But if markets are conceived as constructed entities, they can be skewed to favor certain participants. This is the most important lesson of the FCC auctions.

Precisely this willingness to skew markets in favor of certain participants explains why, despite the failure to implement public policy, the FCC auctions were, as one participant noted, "a huge success for the auction theorists involved."[31] One of the most interesting upshots of the auctions was the development of companies—with many of the key participant game theorists taken on as partners—devoted to the construction of markets.[32]

As Alvin Roth has noted, the FCC auctions opened up "a new way for game theorists to earn their livings, as consulting engineers for the market economy."[33]

It would be a milestone in the history of economics: economists had quickly responded to rapidly changing events initiated by the government, and positioned themselves as the premier experts in market construction. No longer would economists take The Market as something to be explained; instead, they use carte blanche to make up markets in a smorgasbord of shapes and flavors.

But even at the very moment of the Bayes-Nash School's greatest public triumph, the effects of the forces we discussed earlier were becoming apparent. No longer would they talk of English auctions generalized to multiple-item settings, but instead of *the* "Milgrom-Wilson auction." In the years that followed, game theorists would take out patents of various market forms and portray them as made-to-order. The following is taken from the website of one prominent market design company:[34]

> For clients with especially complex problems, we provide customized auction designs that ensure achieving optimal solutions. Complexity raises the stakes in auction design: poor designs leave money on the table or result in inefficient allocations. ... The Milgrom Assignment Auction represents the next generation of multi-product auctions. ... Auctionomics' software is based on Milgrom's innovations in game theory. ... The table below compares the features of clock auctions, traditional sealed-bid auctions, and the Milgrom Assignment Auction [table 15.1]:

Table 15.1 AUCTIONOMICS' PRODUCT COMPARISONS

	Milgrom Assignment	Traditional Sealed Bids	Clock Auctions
Adapted to multiple products	√ Yes	√ Yes	√ Yes
Easy to use with simple substitutions	√ Yes	X No	X No
Allow general expression of substitutes	√ Yes	X No	√ Yes
Finishes instantly	√ Yes	√ Yes	X No
Finds exact market Clearing Prices	√ Yes	√ Yes	X No
Support swaps	√ Yes	X No	X No

With so much at stake, why purchase off-the-rack (or, heaven forbid, wear a hand-me-down) when you can order tailor-made? While Experimentalists had come to this position much earlier, they were still largely underground. It was the (perceived) work of the game theorists in the FCC auctions that changed the self-understanding of the profession as a whole.[35]

[16]

PRIVATE INTELLECTUALS AND PUBLIC PERPLEXITY

The TARP

During the 1990s, market designers had begun to respond to rapid developments in transmutation of actual markets by demonstrating they could skew markets in favor of certain participants. It required them to walk a tightrope: markets would be preferred to bureaucracies on the grounds of transparency and efficiency; yet, only market designers could truly understand their setup and operation. They were so successful in negotiating this treacherous terrain that they were able to take their business model worldwide. Soon thereafter, they participated in auctions in France, Australia, Canada, Mexico, Austria, and the UK, for items ranging from electricity generation capacity, to greenhouse gas emission permits, to railroads. But these economists' amour propre would be seriously challenged by the onset of the global economic crisis, to which we now turn.

Obviously, this book cannot address the crisis as a whole; we only seek to recount a single, limited incident in a long, sad litany.[1] Specifically, let us examine the circumstances surrounding the promotion (and subsequent demise) of the idea that the government could deliver the United States from financial calamity by devising an auction to remove "toxic assets" from the balance sheets of large banks. We have already discussed in chapter 10 how the government changed mortgage markets, leading to the emergence of mortgage-backed securities, and, in turn, to a bevy of toxic assets. Most relevant from the present perspective was that the role of volunteer hazardous materials team was to be played by market designers. Curiously, these theorists were called in to assist with the justification and passage of the U.S. Troubled Asset Relief Program (TARP) in the confusion of late 2008.[2]

The plan to conduct an auction of toxic assets originated in the immediate aftermath of the March 2008 Bear Stearns collapse, and from the conviction among market participants and some Treasury and Fed staff that it would be wise to have a plan to "pull off the shelf" in the case of another Bear Stearns–type emergency.[3] Following several rounds of discussion between staff at the Treasury and the New York and Washington Feds, Neel Kashkari (assistant secretary for financial stability at the U.S. Treasury) and Phillip Swagel (assistant secretary for economic policy at the U.S. Treasury) drafted a memo entitled "Break the Glass: Bank Recapitalization Plan."[4] In this memo, Kashkari and Swagel identified alternative emergency measures, argued in favor of using asset auctions to remove mortgage-related assets from bank balance sheets, and set forth a timeline for completing the asset purchases. Treasury Secretary Henry Paulson would eventually second their judgment to purchase on ideological grounds,[5] but at that juncture he essentially ordered that the plan be set aside.

So when the emergency did eventually arrive, following the September collapse of Lehman Brothers, breaking the glass was something Paulson and Federal Reserve Chairman Ben Bernanke attempted to do. They began to make the rounds to convince members of Congress of the need for an emergency asset purchase plan, solicited an auction plan from the New York Fed, and approached academic market designers to fill in the details.[6] But they almost immediately began to encounter difficulties.

Bernanke gave a performance at Congress for which he was "much ridiculed": during a hearing on the impending asset purchase plan, Bernanke laid out a means to buy troubled assets from banks at "close to the hold-to-maturity price," a slippery magnitude that was highly disputable, but certainly meant paying prices much higher than currently prevailed on asset markets.[7] Serious criticisms immediately surfaced: Doesn't this purchase plan just boil down to giving Wall Street a subsidy? Then why bother with the circumlocutions? Given the nature of the emergency, was it realistic to believe that a relatively small asset purchase plan would do the job? While these objections were gaining intensity in the public sphere, the Treasury worked behind closed doors to craft the original "break the glass" memo into a legislative proposal. The initial effort, which totaled only about 2 and a half pages, was viewed by many as so insubstantial as to be insulting; the House voted down the initial bill based on the proposal. Clearly the plan was in jeopardy.

It was in this context—with skepticism about the asset auctions abounding and financial disaster looming—that market designers assumed a public role in the debate over TARP. Market designers soon found themselves in the public spotlight when Bernanke and Swagel referred to market designers' expertise when fielding concerns about the prices to be paid in their plan,[8] and in short order two of the academic market designers approached by the Treasury, Lawrence Ausubel and Peter Cramton, emerged publicly to defend the legitimacy of the asset purchase plan.

What is interesting about this defense is how market designers portrayed the virtue of their plan. They claimed they could design an auction that would improve upon the Treasury's approach in the sense of establishing lower "competitive market prices," a prospect that did not sound very salubrious from the standpoint of

Bernanke, Paulson, and the bevy of Wall Street lobbyists who had already gone on the record with their concerns about the consequences of driving prices too low.[9] If these microeconomists were to be politically useful, they needed to get on the same page as the Treasury officials and the Fed. Ausubel and Cramton responded by creatively interpreting "competitive market prices" to mean prices that were "reasonably close to value," by which they meant basically the same thing as Bernanke's "hold-to-maturity" prices.[10] The plan purported to allow for the Treasury to manipulate its demand for securities, thereby manipulating the price paid, while preserving the ability to claim that the prices paid were still "market" prices—at least in some sense, an intention that has been subsequently acknowledged:

> A concern of many at the Treasury was that the reverse auctions would indicate prices for MBSs [mortgage-backed securities] so low as to appear to make other companies appear to be insolvent if their balance sheets were revalued to the auction results. This could easily be handled within the reverse auction framework, however ... we could experiment with the share of each security to bid on; the more we purchased, the higher, presumably, would be the price that resulted.[11]

The claim that, armed with the right technique, the Treasury could in effect "go the market one better," while not baldly implausible, did look like they were claiming that the circle could, in fact, be squared: the government could pay greater than market prices, and yet the act of doing so could be rendered "transparent" by the notional market setup. But to the extent that it was possible to ignore this little detail, that would pave the way toward accepting the Treasury's position: issues ranging from executive compensation, to reform of the structural composition of the financial sector, to direct banishment of certain formats of derivatives immediately fell by the wayside. At a time when the most publicly visible economists were arguing against the TARP, the endorsement of these market designers was surely powerful.[12]

According to the market designers, if you understood the crisis from the correct microeconomic perspective you would come to realize how necessary their intervention was. These market designers (of the Bayes-Nash School) argued they could design markets that efficiently aggregate information, and thereby assist market participants to discover the true value of items being sold. They claimed the crisis stemmed from an absence of liquidity, not—crucially—pervasive insolvency. In their frame, banks possessed a variety of assets, some worthless but most others pretty valuable, and it was the inability to distinguish between the two that caused the crisis. By purchasing these assets, the government would reestablish liquidity, not merely by removing toxic assets from the banks' balance sheets but also by releasing information that would establish the assets' true values.

One immediate consequence of this view was that the imposing magnitude of the toxic asset problem was not necessarily worrisome, nor was the possibility that the TARP program would be unable to remove the vast majority of the toxic assets from banks' balance sheets:

> The "losers" are not left high and dry. By determining the market clearing price, the auction increases liquidity. . . . The auction has effectively aggregated information about the security's value. This price information is the essential ingredient needed to restore the secondary market for mortgage backed securities.[13]

What mattered, they insisted, was "information": information would summon forth funds from private actors, thereby thawing frozen secondary markets. The basis for this claim was that the assets to be purchased had a true, objective value that was the same for all bidders, or in the argot of game theory, "common" valued. According to conventional theory, one should expect in such cases that purchasers of such assets (or *sellers* of such assets, in the case of a procurement auction, which in this case was being proposed) should misjudge this objective value, resulting in a kind of undesirable behavior called the "winner's curse." Market designers believed they could mitigate such problems by designing markets that efficiently aggregate information, and thereby assist market participants to discover the true value of items being sold.

Although one way of reading the market designers' argument was that one should generally trust existing markets to do the best job of aggregating information about assets, because the assets were "common value," it was necessary for a suitably trained economist to provide a helping visible hand. In circumstances like these, with the largest financial companies in the nation on the verge of default, the stakes were dangerously high. This made their participation all the more crucial.

In a crisis, especially important was the speed of deployment. Unfortunately, the market designers responded to the Treasury's call for assistance by submitting widely incompatible designs for the auctions, necessitating the Treasury to decide between the rival analyses.[14] For all practical purposes, it was the FCC auctions redux. Complicating matters was that, from the perspective of the Treasury, one could not tell on paper what the best auction form was.[15] For example, one dispute broke out over whether to run an "open" or "sealed bid" auction; historically, that had been one of the most basic issues that market designers had previously grappled with. Which version was to be preferred was supposed to turn on which auction did the best job of aggregating information, but theory provided neither guidance about which format was better nor guidance about whether either form would bring new bits of useful information into the market.

While their early public statements did take care to portray their auctions as market-like, Ausubel and Cramton tended to emphasize to the Treasury how their clock auction improved upon other designs:

> A security's value is closely related to its "hold to maturity value," which is roughly the same for each bidder. Each bidder has an estimate of this value, but the true value is unknown. The dynamic auction, by revealing market supply as the price declines, lets the bidders condition their bids on the aggregate market information. As a result, common-value uncertainty is reduced and bidders will be comfortable bidding more aggressively without falling prey to the winner's curse—the tendency in a procurement setting of naïve sellers to sell at prices below true value.... A principal benefit of the clock auction is the inherent price-discovery feedback mechanism that is absent in any sealed-bid auction format. Specifically, as the auction progresses, participants learn how the aggregate demand changes with price, which allows bidders to update their own strategies and avoid the winner's curse.... Efficiency in the clock auction always exceeded 97%.[16]

In holding that the value of the toxic assets was "roughly the same for each bidder," this passage corresponds to the point made earlier, that market designers viewed the toxic assets as "common" valued, and that such cases posed for the market designer the task of figuring out how to aggregate information. It also makes explicit the mandate of Bernanke's warning to avoid purchasing assets at too-low prices (although market designers offered a different rationale for doing so—avoiding the "winner's curse").

But what is most notable about this passage is that it advocates a specific type of auction—a clock auction—and does so on the basis of its ability to avoid the "winner's curse," as evident by its demonstrably superior "efficiency." The reason this claim is so notable is that it is incoherent on its own terms: it only makes sense to attribute "97% efficiency" in case of private value auctions, where bidders value assets idiosyncratically.[17] If toxic assets are common valued, meaning that all bidders value the assets identically, then all distributions are efficient *by definition*, and therefore the efficiency criterion is useless, or at best, irrelevant.[18]

Above all, this illustrated the resources at the disposal of the market designers, and the lengths to which they were willing to go: they linked avoiding the winner's curse—that is, making sure banks receive enough money from the government—with "efficiency," a goal seemingly more in line with the public interest. To make the case, all they needed to do was get the audience—in this case, the Treasury—to forget the distinction between private and common value auctions—usually presented as lessons one and two in any auction theory text.

But perhaps the point of the exercise was never to get the particulars of the economics justification correct, and instead to get the Treasury to purchase their "clock auction." Sifting through all the coverage of the TARP plan, one comes across an acute observation made by a *Newsweek* reporter:

> [Ausubel and Cramton] hope to convince officials that not only does a reverse auction work, but, in the event the Treasury conducts one, to run it off their patented software platform. . . . Ausubel and Cramton own two auction-services companies, Power Auctions and Market Design, each of which handle

the back end of auctions for companies and foreign governments. They've already helped the French government sell electricity off its grid and Dutch energy companies auction off natural gas.[19]

In fact, Power Auctions and Market Design held the patents for the stipulated clock auction. But the presentation delivered by Ausubel and Cramton for the Treasury listed several additional "typical auction related activities"—namely, product design; definition of detailed auction rules; auction software specification, development, and testing; bidder training; establishment of an auction "war room"; operation of auction; post-auction reports on success of auction and possible improvements for future auctions—for which Power Auctions and Market Design could provide assistance.[20]

Since the days of the FCC auctions, market designers had made all sorts of fantastic claims for their newfangled markets: They can reverse global warming! Improve access to health care! Redress racial and gender discrimination, without committing "reverse discrimination!" Even achieve "free lunch redistribution!" That is, this was so long as you hire their firms to build them to your exacting specifications (after all, "details matter"). They almost always directed the pitch at cash-strapped governments, urging them in particular to sell off public assets to private oligopolistic concerns; in the case of toxic asset auctions, one need only invert the logic. In this case, the government was exhorted to avoid paying too little for assets of dubious provenance. Just as with the FCC auctions, pecuniary interests might have lent confusion to and exacerbated disputes over auction forms. But by now the designers had come to realize that markets could be *patentable*, intensifying yet obscuring the motivations behind the disputes. In stressing how difficult it was to run their *own* market, the market designers did present a compelling case for their ongoing retention; yet it amounted to one more example of how participation as a market designer requires one to withhold trust in the operation of existing markets.

The dispute over auction forms raised a second and more serious problem: there was no good reason to believe that the auctions would do what the market designers had said they would—namely, summon a chain of events that would eventually bring the economy out of crisis by aggregating dispersed information. Some of the Treasury staff became increasingly nervous about performance, regarding the auction design process as a "science experiment" run amok: market designers had always insisted that the performance of the auctions was sensitive to even seemingly minor changes in rules, and yet they could not even agree about how rule changes would affect performance. They wanted to implement the alternative auction forms and use the first set of auctions as trial runs, a prospect that surely failed to inspire confidence. And this in the midst of a collapsing world economy.

Meanwhile, markets themselves had turned against the TARP plan. Things initially had started out well for the Treasury. The first announcement of the toxic asset purchase plan led immediately (on September 18) to a gain on the Dow of 410 points, followed by another 369 points the very next day. Paulson observed that the Treasury's plan had "acted like a tonic to the markets."[21] Unfortunately, matters went from bad to worse, to catastrophic over the course of the next two weeks—at least if one trusted the judgment of markets. The Dow plummeted, and credit markets remained frozen. While it was tempting to attribute the declines to the initial failure to pass TARP, its passage on Friday, October 3, made this a difficult position to maintain, since the declines continued unabated.

When the declines resumed the following Monday and spread across the world, Paulson interpreted financial markets as having judged that "TARP would not provide a quick enough fix."[22] But by then, the handwriting was on the wall: Bernanke and various Treasury staff had been for at least a week expressing doubts that the asset purchase program would work; Paulson himself intimated to President Bush that Treasury would probably need to purchase equity in the banks on October 1, *two days before TARP's passage*.[23] On October 13, Paulson informed the CEOs of Citigroup, Wells Fargo, JP Morgan, Bank of America, Merrill Lynch, Goldman Sachs, Morgan Stanley, Bank of New York Mellon, and State Street Bank that the Treasury now intended to emphasize capital injections— and he instructed these nine banks to accept them.[24] By the end of October, Paulson cancelled the auctions and instructed his staff to concentrate on capital injections instead.[25]

When markets judged the prospective market-based program to be faulty, the Treasury heeded the markets, not the economists. The market designers responded to the Treasury's aboutface in emphasis by insisting that there was no good reason the Treasury could not use auctions to purchase bank shares in addition to toxic assets,[26] a position they maintained until the Treasury made it clear it had no intention to seek release of any additional TARP funds, thereby foreclosing any prospect for using auctions (at least for the remainder of the Bush administration).

Once that happened, things turned ugly: the market designers themselves became some of the fiercest critics of TARP. In an interview for NPR, Ausubel complained, "Instead of conducting transparent auctions, the Treasury is going to instead distribute suitcases of cash"; for Cramton, "It really is moving down the path to crony capitalism, in my mind, where the government is picking winners and losers in a nontransparent way." This turnabout in support for Treasury bailouts was easy to pull off because both the market designers and the anti-TARP petitioners now claimed to have shared very similar assumptions about the economic role of government.

The lesson market designers wanted to draw from the crisis is that the public should trust both The Market *and* the economics profession to rescue them from economic disaster.[27] The rise of market design had already threatened to reveal the contradiction between trusting markets and trusting economists. But the global economic crisis made this position impossible to maintain in the long run; nevertheless, the market designers had proved immensely useful in the sphere of short-term politics. Their participation required them to extol the virtues of The Market, but it also required them to identify ways it had broken down, necessitating their expertise.

The central rationale for employing the market designers was precisely that it held the promise of skewing markets in such a way as to benefit banks holding "toxic assets"; yet it was on precisely this point that they had to promote confusion, lest they be identified with efforts to subsidize the banks. The willingness to engage in such obfuscation was quickly recognized by the Treasury as the market designers' primary virtue, and immediately put to use.

[17]
ARTIFICIAL IGNORANCE

[This new] bio prompts us to ask ourselves why we seem to require of our art an ironic distance from deep convictions or desperate questions, so that contemporary writers have to either make jokes of them or else try to work them in under cover of some formal trick like intertextual quotation or incongruous juxtaposition, sticking the really urgent stuff inside asterisks as part of some multivalent defamiliarization-flourish or some such shit. Part of the explanation for our own lit's thematic poverty obviously includes our century and situation.

David Foster Wallace, "Joseph Frank's Dostoevsky," 1996

And when he occupies a college,
Truth is replaced by Useful Knowledge;
   He pays particular
Attention to Commercial Thought,
Public Relations, Hygiene, Sport,
   In his curricula.

Athletic, extrovert and crude,
For him, to work in solitude
   Is the offence,
The goal a populous Nirvana:
His shield bears this device: *Mens sana*
   Qui mal y pense.

W. H. Auden, Under Which Lyre

In this book we have taken you through the previous seventy-odd years of economics, through computationalism, the crisis, cybernetics, state space formalisms, and much else besides. Following along with us through this history has required some patience and abundant stamina, since we have proposed some unusual topics around which to organize a discussion of the peregrinations of modern economics. Within it all is strung a harrowing misgiving: What did the advent of information have to do with enlightenment in economics?

Obviously we cannot generalize about every economist on the planet. What does this history teach us about the state of orthodox economics in the United States and Europe in the twenty-first century that we could not learn, say, by auditing a graduate course in an economics department? To sharpen our history's denouement, let us compare it with the messages concerning the profession's historical self-understanding, crafted in the halls of today's economics programs and beamed at economic tyros. But where to find such an exemplar amid a profession so unrepentantly hostile to historical self-examination?

What we are after are not stories that practitioners of a discipline tell about *themselves* at ceremonial occasions, the kind of "history" composed by a triumphant graybeard in a mood of reminiscence, possibly with the name "Hayek" or "Keynes" dropped here and there. Instead, we are interested in the views of someone young, preferably fresh out of graduate school, who has relatively little invested in pursuing a specific research program, able to articulate a general viewpoint about recent trends in the profession, but perceptive enough to not dismiss out of hand the notion that the economics profession might actually have something to answer for.

Clearly this kind of person, and this kind of work, would be more easily found in the blogosphere than in scholarly journals like the *American Economic Review* (looking in *History of Political Economy* is a nonstarter). And so we opt to point our browser at *Noahpinion*, the economics blog of Noah Smith, ranked as one of the top twenty-five most influential economics blogs in 2013.[1]

Smith took it as his duty and mission to challenge the unprecedented enmity directed at the economics profession that was suffusing the blogosphere in the wake of the worldwide financial crisis; he responded by drawing attention to what he believes to be praiseworthy recent developments. Not hiding behind impenetrable mathematics and jargon, Smith takes his argument directly to the public. The title of one of his posts accurately conveys his central point: "Economists used to be the priests of free markets—now they're just a bunch of engineers." According to Smith's understanding of the profession, most economists are prone to focus on small, solvable problems, and uninterested in making grand sweeping contributions to policy:

> I have the vague sense that if you were an idealistic, brilliant young libertarian in the 1960s and '70s, you might naturally dream of growing up to be an economist. You might watch a rousing speech by Milton Friedman, and you might imagine that one day you, too, would use the power of logic and rationality and mathematics to ward off the insanity of socialism. Well, America still has some idealistic, brilliant young libertarians, and some of them probably still dream of becoming economists. But now they will be in the minority. They will be joined by quite a few—maybe more—idealistic brilliant young liberals, who recognize the power of markets but also want to figure out how to fix things when markets go wrong. And they will also be joined by quite a few brilliant engineers, for whom political ideals take a back seat to the solving of practical, real-world problems. Econ isn't what it used to be.[2]

It should come as no surprise to discover that Smith elevates market design as the best example of this new form of economics:

> Auctions are one of those situations in which the "agents" are close to perfectly rational–just the type of case that the theorists of decades past liked to sit around and theorize about. This theory worked. And what works, makes money.[3]

To engage in a line-by-line refutation of a blog post, such as this one, would be absurd. Instead, we address it as a representative expression of the public face of the orthodoxy, in order to subject its confident attitudes to critical scrutiny.[4]

Fortified by these chapters, we can now perceive that modern economists live in a topsy-turvy world, where almost everything they believe about themselves is false. Contemporary neoclassical agents are not the epistemic whiz kids of the days of yore—far from it. Yet the orthodoxy has not left all theoretical commitments behind in favor of a judicious pragmatism: their commitments follow a specific path and a curious logic, as we have repeatedly argued. This trajectory has had little to do with any simplistic notions that the theory flat-out "worked": this is the lesson of the last two chapters. Fixing markets with more markets is just another way that neoliberals have of never having to say they're sorry. And the idea that money validates Truth is about as pure a Milton Friedman–style doctrine as one might ever encounter. The profession has become *more*, not less neoliberal; and yet economists are so lost in cosseted political fantasy that it seems they aren't even aware of it. Consequently, they can't see what is plain to many who have suffered through the global economic crisis: trusting The Market to save us may involve withholding trust from the economists who preach its supposed virtues.

It is not only the economists who have been oblivious to trends in the recent history of orthodox economics. Evidently, historians of science these days are faring no better:

> Some economists have come to argue that a prime determinant of bubbles is the amount of disagreement, or "dispersion of opinion," among investors about an asset's value. Economists have explored various factors that might permit such significant disagreements, notably including investors' overconfidence and limited attention. A few economists have even begun to think about investor disagreement in terms of differences in *knowledge* and to borrow theoretical insights from the history and philosophy of science. Looking to *disagreement*, it seems, may allow for new (and agreeable) collaborations between humanistic and social scientific interpretations of economic life.[5]

Economists have, of course, done no such thing. As we have shown, economists now believe knowledge to be right in their wheelhouse, after a little prior help from their natural science comrades. Hence, they would never feel compelled to borrow from the humanities— that would be squatting with the untouchables.[6] And as for their openness to the history and philosophy of science, well, one might just as cogently suggest that their contempt for philosophy is what eminently qualifies professional economists as engineers of the human soul. Only someone completely unfamiliar with the theoretical traditions we have covered in this book could make so absurd a claim.[7]

How have we arrived at this impasse, with artificial ignorance ingrained at every turn? How have so many smart people so misconstrued the trend of modern economics? How can economists pride themselves as wizards of information, and yet be so woodenly obtuse? We cannot do justice to those large issues here, and thus we must leave the larger question to those who come after us. However, we wish to reiterate the observation that the Neoliberal Thought Collective is quite happy to have the masses mired in artificial ignorance, since that merely greases the wheels of The Market, that for which there is no greater intelligence.

THE BURDENS OF EPISTEMIC PRIVILEGE

By now, we can appreciate that the ambition to engineer an economy has been intertwined with the quest to find a place for information in economics. "Information" has endured a curious career in economics: as it became increasingly remote from the human, humans were enlisted in the operation of information processors, and this operation came to be determined by those designing markets. In the first instance, game theorists ascribed incredible powers of ratiocination to agents.[8] This would inevitably raise questions about the role economists would play in the larger scheme of things, to which economists responded by stepping back from the precipice and insisting that agents were prone to fall victim to the "winner's curse," and therefore required expert assistance after all. It was important that market designers not portray agents as *too* smart.

Experimentalists took this lesson to heart, and portrayed agents as less than incredibly rational and, for certain purposes, as stupid as it is possible for a person to be. In terms of the historical trajectory discussed in chapter 10, the profession came to hold that its task was to build markets in such a way that agent cognition should be irrelevant to their successful operation. Agents would be folded into the person–machine system, no longer deemed capable of understanding why they made the decisions that they do. Think of their predicament as Artificial Ignorance. Such understanding would be reserved for the economist alone.

One cannot make sense of this development absent an understanding of the increasing influence of neoliberalism. Although the language of information processors and person–machine systems does seem to conjure the image of the engineer occupied by technical matters, the orthodoxy would often portray The Market as not merely an information processor but also an *omnipotent* processor of information, That Than Which No Greater Can Be Conceived. While it may make sense to describe markets as information processors, it makes little sense to portray them as omnipotent processors, at least from the standpoint of the abstract theory of computation.

Here, one should understand the orthodoxy as coming round to the view of Hayek, who himself came to espouse the principle that markets could "know" things that agents could not. Similarly, the ability of the economic actor to arrive at knowledge independent of the "smart market" would be denied: MPS member Vernon Smith wrote, "There is a puzzle as to the processes whereby our brains have [market exchange] and other skills so deeply hidden from our calculating self-aware minds."[9] Neoliberalism influenced the way computational themes would enter economics: the agent would become one small cog in the grand market mechanism.[10] The central point obscured by counterposing market engineering with policy advocacy is that neoliberal theory was a central influence on the development of market design.

Consequently, knowledge no longer looks like it did in the Enlightenment roots of political economy. What happened to the Kantian subject, able to reason for herself, autonomous, and hence an end in herself? Economists' fascination with information has inadvertently debased their treatment of knowledge—first, for the agent and then, ultimately, for the economists themselves. Now all we have left is *information*. It was a seemingly technical notion that, reified, was then progressively removed from the grip of the agent who, in turn, would be denied anything that could reasonably be signified as "understanding" or even "thought." This *neoliberal* subject was banished from the realm of ends, denied any optimality that makes sense, fated to slave away on a supremely complex calculation, churning through a subroutine, Truth always eluding its grasp.

Forget postmodernism: this is the contemporary Death of the Author, the ultimate debasement of freedom in the name of freedom to choose.

In the introduction we asked: "Why must an economics of knowledge end up propounding a doctrine that economic agents should be ignorant?" The tenacious reader will notice that we have been working our way toward a twofold answer. First, the imperative to design person–machine systems necessitated that agents be conceived as relatively stupid, at least when it came to economic rationality, and economists as sufficiently smart enough to understand how to orchestrate the amalgam of persons, machines, and markets into a knowledge producing whole. In their models, when agents are folded into person–machine systems, then the market-designing economist alone is granted the god's-eye knowledge to understand and shape the operation of the market—otherwise, the designer has no role to play.

This epistemic privilege does not come cheaply. In practice, market designers have managed to participate in designing markets not by embodying neutral expert knowledge but, instead, by promising to skew markets in favor of specific actors. In the FCC case, these actors were the telecoms; in the TARP case, they were the banks. But this required market designers to promote confusion about how markets can be skewed in practice, by publicly conflating markets with The Market, by referring to "market prices" and the transparency of "the market" when compared with supposedly opaque bureaucracies in both the FCC and TARP cases, while at the same time acknowledging that subtle differences in market structures can lead to radically different outcomes.

The practical requirements of suppressing market diversity has by now transmogrified into professional wisdom, nicely channeled by John McMillan:

> In addition to markets, there is also the market, an abstraction as in "the market economy" or "the free market" or "the market system." The abstract market arises from the interaction of many actual markets.[11]

The current generation of economists apparently remains untroubled by this strange portrayal of a pristine abstraction as an emergent property of "actual markets"; it smacks of an organicist transformation of Quantity into Quality. It is probably closer to the mark to say that their training has prepared them to waver almost effortlessly between the two contradictory positions: "as economists find more and more theories that predict how markets actually behave, they've moved beyond the policy realm and into the realm of engineering."[12] In other words, there is one image of the grand Walrasian general market for political consumption, and a different image of a collection of diverse boutique markets operating with differential effects for the clients of the business of market design.

What history can do is reveal that these two dogmas are contradictory; no amount of wishful thinking can dispel the dilemma. The market cannot simultaneously be a grand monolithic oracle of information and a motley jumble of variously wonky information prosthetics. Not only do the imaginary agents in contemporary models display deficiencies of epistemic capacity, but also the modern economic orthodoxy has taken to degrading the knowledge of the real public, as revealed in our narratives of the spectrum and TARP auctions. The business models that create the space for economists' participation in the modern transformation of market structures dictate that they promote confusion about how markets can be effectively skewed in favor of some players and against the interests of others; consequently, their public proclamations have made real people stupider about actual markets.

As Maureen Tkacik put it in commenting upon the present crisis, the general public responded to events by wanting to feel less stupid; but was then shocked by how stupid those in positions of authority had appeared to have been.[13] Perhaps some of this impression is due to the public role played by the economics profession. Hence, the public role played by market design economists is the second way agents were made ignorant: for economists to participate as market designers, they would have to promote confusion about the very nature and identity of markets.

Bearing that in mind, recall that one of the central commandments of neoliberalism was that The Market alone can arrive at Truth. But how does the manipulability of markets affect the belief that markets arrive at Truth for all—the dream of a well-functioning knowledge economy? If markets are truly diverse, and they can be made to order, then why should anyone expect a priori that they would reliably reveal the Truth?

NOTES

Chapter 1

1. For the background of Knight's quest, see Emmett 1998, one of the truly great papers in the annals of the history of economic thought.
2. Knight 1940, p. 4.
3. As we write this, another example of dissention has roiled the blogs: Romer 2015.
4. See, however, Rizvi 1994, 1998.
5. See Gul and Pesendorfer 2008.
6. Gentzkow and Shapiro 2008, p. 140, based on the work of Milgrom and Roberts 1986.
7. Rodrik 2014, p. 189.
8. The important role of the Mont Pèlerin Society as the early incubator of neoliberalism in our narrative is discussed in chapter 4 and in Mirowski and Plehwe 2009.
9. Robert Barro, in www.economist.com/blogs/freexchange/lucas_roundtable.
10. Hence we strenuously disagree with the historian Samuli Leppälä, when he writes (2015, p. 276), "justified true belief still provides a good working definition of knowledge for economists." Most of the protagonists in our narrative would have little idea of what that notion would even imply for their models. An interesting sidelight to this issue is the movement within philosophy to dispense with the standard Philosophy 101 definition of "justified true belief" in favor of treating "knowledge" as an explanatory primitive concept. On this, see

NOTES

McGlynn 2014. It has occurred to us that this parallels the historical reification of the information concept, described herein.

11. Foucault 2011, February 1984, first hour. See also Folkers 2016.
12. "Tales of Ratiocination," this section's title, is what Edgar Allen Poe called what later were praised as the earliest "detective stories."
13. Horning 2012.
14. Michael Chabon, at: http://www.nybooks.com/articles/archives/2013/nov/07/thomas-pynchon-crying-september-11/.
15. Amazingly, Horning does not mention the modern master and parodist of such novels, Thomas Pynchon.
16. Horning 2012.
17. Not unexpectedly, the same dynamic also shows up in modern healthcare. Joseph Dumit (2012, p. 1) defines the modern conception of health as "always being at risk—and never knowing enough about what you should be doing."
18. See, for detail, Mirowski 2011, chap. 2.
19. http://kingsreview.co.uk/magazine/theblog/2015/04/08/neoliberalism-1979-2008/. The *reference then leads to a clarion call to reorganize the social sciences: "So social scientists should devote a small palace guard to settled subjects and redeploy most of their forces to new fields like social neuroscience, behavioral economics, evolutionary psychology and social epigenetics, most of which, not coincidentally, lie at the intersection of the natural and social sciences. Behavioral economics, for example, has used psychology to radically reshape classical economics." See http://www.nytimes.com/2013/07/21/opinion/sunday/lets-shake-up-the-social-sciences.html?_r=2.

    See also "After the Crash, Can Biologists Fix Economics?" *New Scientist*, July 22, 2015, pp. 38–41.
20. The ways in which the Bourbakist school of mathematical abstraction influenced mid-century neoclassical microeconomics is described in Mirowski and Weintraub 1994.
21. Veblen 1930, p. 73
22. See, for instance, Erickson 2010.
23. Boulding 1966, p. 1.
24. See Klaes and Sent 2005.
25. See, especially, Purcell 1973; Heyck 2012.
26. See Pareto 1981, p. 41.
27. These factors were first explored in detail in Mirowski 2002. Because that book stresses the military influences, we do not supply further independent documentation of that aspect of the narrative in this volume.
28. Heyck 2012, p. 100; emphasis added.
29. Erickson et al. 2013, pp. 32, 157.
30. See Hands 2015; Davis 2006, 2008.

NOTES

Chapter 2

1. Eric Beinhocker, INET, Oxford University, blurb for Hidalgo 2015.
2. See the authors in Davies and Gregersen 2011 for instance, or Lloyd 2006.
3. See Gleick 2011; Agar 2012, chap. 16; and Kline 2015.
4. William Shakespeare, *Henry VI*, Part Three, Act III, Scene ii, lines 191–193.
5. Dang et al. 2012. This is doubly ironic, given that Gorton was partially responsible for the promulgation of the financial instruments that blew up AIG in the financial crisis. See Mirowski 2013, pp. 209–210.
6. See Hollinger 1997, pp. 345–346; emphasis added.
7. This trend was first described in the underappreciated work of Esther-Mirjam Sent (1998).
8. This problem of reflexivity began to surface in the 1950s, in a number of social sciences, about the possibility of "self-fulfilling prophecies." On this, see Hands 1990. Interestingly, Norbert Wiener refused to comment on this literature when asked (Kline 2015, p. 135).
9. This ignores a rather sad reverse phenomenon, that once some novel intellectual trend becomes apparent in modern economics, someone somewhere attempts to read its content back into the classical economists, usually Adam Smith, seeking to insist that economists really knew about it all along. For an example explicitly attempting this retrospective whitewash with regard to "information," see Prendergast 2007.
10. See https://afinetheorem.wordpress.com/2015/03/06/the-contributions-of-the-economics-of-information-to-twentieth-century-economics-j-stiglitz-2000/.
11. The Arrow quote is taken from Colander et al. 2004, p. 292.
12. See Daniel Klein 2005, p. 105. The next sentence reads: "In my old-fashioned positivism, concepts have meaning only in the context of a model (which may be very general), and I can't think of one which will accommodate this distinction."
13. Hidalgo 2015, p. 79.
14. Boulding 1968, p. 142. Of course, other economists insisted upon similar distinctions; one such theorist, George Shackle, was also banished from orthodoxy (Basili and Zappia 2009).
15. We would point here to Babe 1994; Machlup and Mansfield 1983; Lamberton 1998; Levine and Lippman 1995. Things have not much improved in the interim, judging by Leppälä 2015.
16. Levine and Lippman 1995, p. xii.
17. Schiller 1988, p. 32.
18. Stiglitz 1985, p. 22.
19. See, for instance, Shapiro and Varian 1999; Varian 2002; Kreps 1997.
20. Twenty-five years on, David Kreps's (1990, p. 578n) warning is still good advice: "The terms of information economics, such as moral hazard, adverse

selection, hidden action, hidden information, signaling, screening and so on are used somewhat differently by different authors, so you must keep your eyes open when you see any of these terms in a book or an article.... As a consumer of the literature, you should pay less attention to these labels and more to the 'rules of the game'—who knows what when, who does what when." The only codicil one might add is to replace "consumer of the literature" with "epistemically challenged member of the economics community," which better captures the repressed paradox.

Chapter 3

1. See Gleick 2010.
2. But consult Mirowski 2002; Gleick 2011; Aspray 1985; Kline 2015; and the useful website http://monoskop.org/Information_theory.
3. See Mirowski 2002, pp. 68–76, for the cryptography; and Gleick 2011, chap. 6, for the broad story of Shannon's notions. The key paper was Shannon 1948.
4. Here are some historians who have made this observation: Mirowski 2002, pp. 72–76; Heims 1991; Kay 2000; Machlup and Mansfield 1983; Kline 2004; Levy 2011. For the student interested in an overview of the main histories of the shape-shifting character of information from Shannon onwards, see the very perceptive paper by Geoghegan 2008 and the website http://monoskop.org/Information_theory.
5. See von Foerster 1952, p. 22; Geoghegan 2008; Shannon 1956.
6. See Kay 2000 and Levy 2011.
7. In Machlup and Mansfield 1983, p. 52.
8. Quastler 1955, p. 2.
9. For the Machlup quotes, see Machlup and Mansfield 1983, pp. 653, 649. For the history of conflation, see Kauppinen 2013.
10. See Solovey 2013, chap. 2; Thomas 2015; and Heyck 2012.
11. Both are covered in detail in Mirowski 2002.

Chapter 4

1. For more on the curious case of the *ersatz* Nobel Prize in Economics, see Mirowski 2015.
2. For some histories of the MPS, see Hartwell 1995; Mirowski and Plehwe 2009; Burgin 2012; Mirowski 2013.
3. The following is based on Mirowski and Plehwe 2009; Mirowski 2013; Mirowski 2002, pp. 232–241. Since 1980, the MPS has become less functionally central to neoliberal organization, which is why we do not extend table 4.1 beyond 1995.

If we also include a few other key MPS members who did not receive Nobels, but played a large role in the early economics of information, such as Fritz Machlup and Henry Manne, then the MPS connection becomes even stronger.
4. There is a wonderful glimpse from the archives of the Samuelson Papers at Duke University that makes this point. In the early 2000s, the journalist David Warsh was writing a popular book claiming a recent grand transformation of economics into a theory of information, with Paul Romer as hero, later published as *Knowledge and the Wealth of Nations* (2006). Samuelson had read a draft, and challenged Warsh's timeline: "There was a long history of interest in the economics of knowledge. Theorists like Fellner, Weizsacker, Kennedy, Arrow.... Substantial book written and edited by Fritz Machlup. (Have a look at that one.) What Romer did was to invent a neat do-it-yourself Tinkertoy" (Paul Samuelson to David Warsh, July 30, 2002, Samuelson Papers, Duke University, Box 76, folder: Warsh).
5. See Burgin 2012; Mirowski and Plehwe 2009; Van Horn et al. 2011; Mirowski 2013; Jones 2012.
6. This is discussed in greater detail in Caldwell 2004; Mirowski 2002, 2007, 2011. Hayek (1982) himself admits the dependence upon brain metaphors.
7. See Foucault 2008, p. 226.
8. Hayek 1960, p. 81.
9. See Plant 2010, p. 67.
10. Hayek 1967, p. 172.
11. Benjamin Constant, cited in Röpke, 1949, p. 28.
12. Posner, quoted in Harcourt 2011, p. 147.
13. Ibid., p. 149.
14. This problem is nicely captured in a post on the website Social Democracy for the 21st Century: "(1) Austrians are deeply divided on the significance and even truth of Hayek's "knowledge problem"; (2) the Austrians cannot get their story straight on what it is that makes rational economic calculation in a planned economy impossible" (March 22, 2013, at http://socialdemocracy21stcentury.blogspot.com/).

Chapter 5

1. For background, see Lavoie 1985; Hayek 1948; Caldwell 2004.
2. See Van Horn and Mirowski 2009; Nik-Khah and Van Horn 2015.
3. The prehistory of neoliberalism growing out of Weberian sociology has been the subject of recent intensive research by Nick Gane. See https://estudiosdelaeconomia.wordpress.com/2014/09/14/is-neoliberalism-weberian-an-interview-with-nicholas-gane/. The Weberian notion of ideal types also materialized in "type theory," described in chapter 8.

4. Hayek 1948, pp. 77–78.
5. See Simon 1968/1981, p. 41.
6. Samuelson 2009, p. 2.
7. Solow 2012.
8. Boettke and O'Donnell 2013, p. 306.
9. Some modern orthodox economists (Myerson 2009; Stiglitz 2000; Reiter 2009) have admitted this but, we think, without appreciating its full significance.

Chapter 6

1. Pride of place goes to Oguz 2010; other useful sources are Kahlil 2002; Lavoie 1985; Boettke and O'Donnell 2013.
2. Hayek 1988, p. 88.
3. It may not be amiss to point out that this structural similarity to Shannon's use of entropy was one reason the Hayekian "first movement" proved so popular well outside the professional precincts of economics, as in artificial intelligence.
4. Hayek 1952/1976, p. v.
5. Ibid., p. 24.
6. On the structure of associationist psychology, see Daston 1978; Mandelbaum 2015.
7. It has not been clear to subsequent commentators just how different, and even opposed, were Polanyi's and Hayek's philosophies of knowledge. This has been occluded by assertions that they both believed in similar notions of "tacit knowledge." On this, see Mirowski 1998; Oguz 2010; Bateira, in Dolfsma and Soete 2006; Butos 2010. Nevertheless, the Hayekian version of unconscious rationality was popularized for non-economists in Gladwell 2005.
8. See Boettke and O'Donnell 2013, p. 314: "Radical ignorance is a significant element of Hayek's thought." But Hayek originally thought of it as a refutation of the sociology of knowledge: "the whole aim of the discipline known under the name of the 'sociology of knowledge' which aims at explaining why people as a result of particular material circumstances hold particular views at particular moments, is fundamentally misconceived" (Hayek 1952/1976, p. 192).
9. Hayek 1978, p. 46. Hayek's main biographer, Bruce Caldwell, seems to misunderstand the significance of this alteration in the approach to knowledge and consciousness in the later Hayek. He suggests "Hayek never directly linked his economics and his theory of the mind" (2004, p. 277); but that impression derives from Caldwell asserting Hayek's first movement persisted throughout his career.

NOTES

10. Hayek 1978, pp. 179, 182, 183; emphasis added.
11. Rumsfeld was himself a member in good standing of the neoliberal thought collective, an avowed acolyte of Milton Friedman. For the quote, see Rumsfeld 2010. It has been reported that Donald Rumsfeld, in a speech at Milton Friedman's 90th birthday party in 2002, which was held by the Bush White House to honor Friedman's legacy, said, "Milton [Friedman] is the embodiment of the truth that ideas have consequences."
12. Hayek 1988, p. 88. In particular, he denounced any idea that information should be freely available: "To conceal and to use superior information for individual or private gain is still regarded as somehow improper—or at least unneighborly" (p. 101).
13. Hayek 1978, p. 183.
14. See, for instance, Diamond 2012.
15. Cowan et al., 2000 propose a similar taxonomy—articulated, unarticulated, and inarticulable knowledge—without citing Hayek, only to reject the third category as "not very interesting for the social sciences" (p. 230). This book shows just how misguided their judgment was.

Chapter 7

1. Some internalist memoirs include Christ 1994; Hildreth 1986; Klein 1991. There is a tendency in those texts to stress the early achievements of Cowles in econometrics, which turned out to be perhaps the least important aspect of its history. If we had to summarize its achievements in order of importance, they would be: (1) introduction of the models of information into American orthodoxy; (2) genesis and promotion of Walrasian general equilibrium as neoclassical American orthodoxy; (3) innovation of the (anti-Keynesian) rational expectations approach to macroeconomics and money; and (4) development of full-information maximum likelihood statistical techniques for estimation of sets of simultaneous equations.
2. See Warsh 1993, p. 64.
3. For details, see Mirowski 2002, pp. 216–220.
4. Leonid Hurwicz to Ruth Schechter, September 12, 1940, Leonid Hurwicz Papers, Perkins Library, Duke University Archives, Box 23, File: Correspondence 1940.
5. Jacob Marschak Papers, UCLA Young Library, Box 91, Folder H.
6. Jacob Marschak to Chancellor Lawrence Kimpton, May 25, 1951, Jacob Marschak Papers, UCLA Young Library, Box 92.
7. Koopmans 1951, p. 3.
8. Arrow 2009, p. 7.
9. See Heyck 2012, 2015; Bessner and Guilhot 2016; Thomas 2015.

10. For documentation of this assertion, see Mirowski 2002, chaps. 4 and 5; Augier and March 2011, esp. pp. 68–89.
11. There have been a number of insightful histories commenting upon economists' encounters with psychologists, but for reasons of concision we shall have to make do here with a couple of bald generalizations, based on Giocoli 2003, to get our story rolling. This was first broached in chapter 1.
12. For a detailed summary, see Mirowski 2002, pp. 370–389. Even there, many important Cowles initiatives are left unexplored.
13. Arrow 1962, p. 614.
14. Koopmans 1957, p. 53.
15. Comments in Thursday afternoon session, Conference on Expectations, Uncertainty and Business Behavior, Pittsburgh, October 27–29, 1955, Box 5 folder 81, Tjalling Koopmans Papers, Sterling Library, Yale University. Note that even though Koopmans was close to von Neumann in this era, he did not entertain the notion that game theory was a better formalism for addressing these questions.
16. See McGuire and Radner 1986, p. viii.
17. See Mirowski 2012,
18. Marschak 1974, vol.1, p. 93.
19. But certainly *not* "the first to develop a systematic theory of the economic value of information," as asserted in his biography in the *New Palgrave* by Roy Radner (see Mirowski 2002, pp. 372–375).
20. Jacob Marschak to John McCarthy, November 27, 1957, Jacob Marschak Papers, UCLA Young Library, Box 94, Folder: M.
21. Jacob Marschak to Howard Raiffa, July 7, 1966, Jacob Marschak Papers, UCLA Young Library, Box 111, Folder: R.
22. Jacob Marschak, "Elements for a Theory of Teams," October 1954, Jacob Marschak Papers, UCLA Young Library, Box 90. Another motivation for team theory was to produce models of command and control for the military, something Marschak acknowledged in print (see Marschak 1974, vol. 2, pp. 64–66).
23. Leonid Hurwicz, "Economic Decision-making Processes ... ," [no date, possibly 1951], Jacob Marschak Papers, UCLA Young Library, Box 91, File: Hurwicz.
24. A case can be made that Hurwicz and his followers initially ignored the twists and turns of Hayekian epistemology covered in chapter 6, only to have the dumbing down of the agent come back to bite the next generation of neoclassical orthodoxy.
25. "I was writing a more or less expository paper on dealing with activity analysis.... [W]hen I used the word, 'decentralization' I thought I should explain what it meant.... But then it struck me that I did not in fact know what we meant by decentralization. That was the beginning of many years

NOTES

of work trying to clarify the concept" (Hurwicz, quoted in Feiwel 1987, pp. 271–272).

26. See Hurwicz 1969, p. 517.
27. Reiter 1977, p. 230.
28. See Simon 1991 and also Sent 2001.
29. "Inquiry on Cowles Commission," Memo from Herbert Simon to Clifford Hildreth, August 2, 1982, Clifford Hildreth Papers, Perkins Library, Duke University Archives, Folder: Correspondence.
30. See Sent 2004 on this issue.
31. Simon 1991, p. 163.
32. Simon 1978, p. 500.
33. For the story of Kramer and Lewis, see Mirowski 2002, pp. 422–432.
34. For the explicit repudiation of the Shannon concept, see Arrow, in McGuire and Radner 1986. For the admission that his models had little to do with cognitive information processing, see Arrow 1984, p. 200. "There is no general way of defining units of information" (Arrow 1996, p. 120). For Arrow's role in suppressing the work of Alain Lewis, see Mirowski 2002, pp. 427–436.
35. Kenneth Arrow, in Colander et al. 2004, pp. 293, 298.
36. Radner 1968, p. 35.
37. The exception is Lee 2015, to which we owe our appreciation of his importance.
38. They were primarily known as the authors of a widely used microeconomics textbook of the 1970s: Quirk and Saposnik 1968.
39. Ames 1981, p. 358.
40. Smith 2008a, p. 230.
41. Reiter 1977, p. 227.
42. Reiter and Mount 1974, 2002.
43. Smith 1991, p. 162. Reiter had moved to Northwestern in 1967.
44. Reiter et al., 1989. See also Lee 2015.
45. See Reiter 2001, p. 271.
46. Reiter and Maroulis 2008, pp. 1399–1400.
47. See Kline 2015, p. 101.
48. See Arrow et al. 1949.
49. See Blackwell 1953; Blackwell and Girshick 1954. The RAND inspiration for the approach was admitted by Blackwell in (DeGroot 1986, p. 47): "My work on the comparison of experiments was stimulated by some work by Bohnenblust, Sherman and Shapley."
50. See Fourcade and Khurana 2013.
51. Eran Shmaya, "David Blackwell," at https://theoryclass.wordpress.com/2010/07/18/david-blackwell/. The shape of the Blackwell legacy is described in chapter 8 this volume.
52. See, for instance, Samuelson 2004.

NOTES

Chapter 8

1. Richard Langlois, quoted in Machlup and Mansfield 1983, p. 586.
2. Dorfman 1960, p. 585. The notion that those devious relativists who thrive in science studies are the only cadre who are susceptible to the perils of reflexivity is one of the sillier arguments made by modern philosophers.
3. Quoted in presentation by Judy Klein, "The Militarized Zone between Theory and Practice in Economics," at www.cigionline.org/sites/default/files/shared/Plenary%20Session%202-Judy%20Klein_0.pdf.
4. Gary Becker, "Age of Human Capital," at http://down.cenet.org.cn/upfile/44/200541910475132.pdf.
5. Akerlof 2002, pp. 411, 413.
6. See Romer 1990; Warsh 2006.
7. See, for instance, Boyle 2000; Jaffe and Lerner 2004.
8. See Gigerenzer and Murray 1987.
9. "History has witnessed the attempt to make probability theory coherent with what was believed to be rational thought, and it has seen efforts to reduce rational thought to probability theory. For instance, what was believed to be rational judicial and economic thought actually determined the way in which probability theory developed mathematically" (Gigerenzer and Murray 1987, p. 137).
10. The historical background to this development is covered in Mirowski 2002, pp. 380–386. A nice introductory analytical treatment from the standpoint of epistemic logic is Fagin et al. 1995.
11. In this latter case, we observe one of the few instances where professional philosophers played a significant role in the development of a notion of knowledge that later became important in economics. The reason this happened was that many of the philosophers in question were also active at RAND in their other capacity as operations researchers. The story begins with Rudolf Carnap (1947) and reaches a high level of development with Saul Kripke (1963).
12. See Mirowski 2002, pp. 380–385.
13. See Fagin et al. 1995.
14. Diamond and Rothschild 1978.
15. Some examples of this literature include Akerlof 1970; Rothschild and Stiglitz 1976; Spence and Zeckhauser 1971 (all reprinted in Diamond and Rothschild 1978). There also was a fair bit of confusion with the modeling strategy of information as thing or commodity in this era.
16. Fagin et al. 1995, p. 32.
17. Michael Rothschild, "Models of Market Organization with Imperfect Information" (1973, in Diamond and Rothschild 1978, p. 479).

18. Ibid., p. 461: "Models of what is usually called disequilibrium behavior do not make sense and cannot serve as reliable guides to further theorizing or policy unless they are consistent and coherent." Here we observe what is usually considered the "Lucas critique" was well established in orthodox microeconomics at that time.
19. This magic trick has been discussed in detail by Davis 2011.
20. There is a suppressed genealogy of this construct rooted in the Weberian sociology of "ideal types." Game theorists seem not all that interested in social science motivations for this mathematical trick.
21. Dekel and Siniscalchi 2014, p. 4. Would that applied game theorists were so circumspect.
22. Myerson 2004, p. 1823.
23. See Taylor 1998; Mirowski 2002, pp. 82–85.
24. See, for instance, http://www2.econ.iastate.edu/tesfatsi/ace.htm.
25. Matthew Rabin, quoted in Colander et al. 2004, p. 141.
26. On this, see Mirowski 1989.
27. See, in particular, Mirowski 2002, pp. 422–436.
28. Smith 2008a, pp. 194–196.
29. Smith 2001, p. 428.
30. See Mirowski 2002, pp. 551–560 for various attempts to draw out the implications.
31. See Sunder 2004.

Chapter 9

1. "The comparative merits of alternative systems are typically being debated under such labels as centralization against decentralization, social control or planning against free markets, or in similar terms. This dichotomy was present in the famous Mises-Hayek-Lange-Lerner controversy concerning the feasibility of socialism. . . . A survey of the literature will show that issues concerning the proper internal structure of businesses and other large organizations involves [sic] similar dichotomies" (Hurwicz 1971, p. 80). For a recollection of this work, see Reiter 2009.
2. He associates what he calls "high modern social science" with "An abiding interest in the means by which systems store, process, and communicate information about themselves and their environments, often expressed through the formal analysis of information" (Heyck 2015, p. 11).
3. "Problems of economic policy may be grouped in two broad classes which may be loosely described as those involving choice of the value of a "parameter" within a given system of economic institutions and those involving choice among institutions. . . . Examples of the second type include the design of "new"

NOTES

economic systems, such as were embodied in the Yugoslav economic reform of 1968" (Mount and Reiter 1974, p. 161).
4. See, for example, Milgrom 2004.
5. The questions are taken from a National Academies report drafted for the express purpose of setting research priorities (Reiter et al. 1989, pp. 284, 285), first described in chapter 7 this volume.
6. Reiter et al. 1989, pp. 286, 304–305.
7. We will review the winner's curse and discuss its significance for the development of market design in chapter 12 this volume.
8. For the moment, we must be deliberately vague about the goals economists believed were appropriate to pursue; suffice it to say that during this crucial transformative period they had begun to acknowledge that the traditional criterion of Pareto optimality was inadequate to the task, for reasons we will begin to elaborate upon in the final section of this chapter.
9. See Lee 2015.
10. The profound changes to real-world markets that had taken place during the 1970s and 1980s forms one of the subjects of the next chapter.
11. Although we tend to forget it now, pioneers in this effort often wore their neoliberal commitments on their sleeve. For example, Charles Plott credited his work in design to Buchanan's "constitutional political economy" (see Lee 2015).
12. Giocoli 2009, p. 204.
13. Giocoli 2003, p. 405.
14. See the textbook histories presented in Angner 2012; Cartwright 2011; Wilkinson and Klaes 2012.
15. For this reason, we find Catherine Herfeld's (2013) attempt to illuminate changes in the economics orthodoxy by examining the career of Jacob Marschak to be unpromising. While Marschak did grapple with information, he did not address markets. We would argue that this is the primary reason his project hit a dead end.
16. It should therefore be unsurprising that Hurwicz's career received little sustained attention at the hands of historians. Weintraub 1991 gives him the most attention; Ingrao and Israel 1990 barely mention him; Niehans 1990 devotes a mere footnote; and Blaug 1996 omits him entirely. Previous attempts to reevaluate Hurwicz's significance include Mirowski 2002; Lee 2006; and Nik-Khah 2005.
17. Maskin was 57 at the time of the prize; Myerson was 56. Only Kenneth Arrow (51), Robert Merton (53), Paul Samuelson (55), and Paul Krugman (55) were younger at the time of their reception of the prize (William Sharpe, Myron Scholes, and James Heckman were all 56). See www.nobelprize.org/nobel_prizes/lists/age.html.

NOTES

18. One former Minnesota PhD student took exception to the attempt to write Hurwicz's earlier work out of the history, but attributed it to previous economists' unfortunate lack of facility with mathematics: "[In] Mr. Hurwicz's first seminal contribution 'Optimality and Informational Efficiency in Resource Allocation Process' (1960) . . . the search for an allocationally and informationally efficient, plus incentive-compatible mechanism started. Because of its mathematical rigor and difficulty, very few economists follow[ed] the study, and it took almost five decades for the Nobel Prize committee to recognize the importance of the study and Mr. Hurwicz's contribution" (Liang-Shing Fan, Letter to editor, *Wall Street Journal*, October 27, 1987). As will become clear, we disagree that lack of mathematical facility was the reason for Hurwicz's late recognition.
19. David Warsh, "The Road to a System that Works (Without Shooting People)," *Economic Principles*. At www.economicprincipals.com/issues/2007.10.21/69.html.
20. Myerson 2009, p. 60.
21. As of this writing (August 11, 2015), there is still no dedicated entry for Hurwicz in the *Palgrave*.
22. In his first published paper on mechanism design, Hurwicz introduced one such mechanism, the "greed process": "Its origin was an old Polish Jewish anecdote about a young man who went to buy a suit. But he had never bought anything before. So his father told him, 'Whatever they ask, always offer half.' So when he was asked, let us say 100 zloty, he said 50 zloty. When the tailor went down to 80 zloty and he retorted 40 zloty. At the end the tailor is really disgusted, wants to get rid of him, and tells him he can have the suit free. The young man then retorts, 'Can I have two pairs of pants?' The 'greed process' is somewhat similar in spirit" (Hurwicz, quoted in Feiwel 1987, p. 271).
23. Hurwicz long remained an enthusiast of cybernetics. See Kline 2015, p. 306n36; Gerovitch 2009. On Hurwicz and cybernetics, see also Mirowski 2002; Lee 2006.
24. "Both [mechanism design theory and control theory] are normative in spirit. That is, they do not accept the status quo, but look for modes of intervention that would bring the system as close to optimality as possible. Thus the mode of intervention is the unknown of the problem. But while rejecting a purely passive attitude toward the working of the system, they also try to avoid the danger of Utopianism by taking into account the constraints to which intervention and its effects are subject. . . . What does seem important is our ability to manufacture such synthetic systems from purely mathematical considerations, rather than having to rely on precedents of observed economies" (Hurwicz 1979, pp. 123–140).

25. Feiwel 1987, p. 273.
26. In that paper, Hurwicz had insisted that because it was equally an issue for all relevant economic systems that it was not worthy of sustained attention, and therefore framing incentive compatibility as a political issue involved a bit of a post-hoc revision.
27. However, see Hurwicz and Thomson 1991.
28. Hurwicz 1972/1986, p. 301.
29. Mount and Reiter 2002, p. 12.
30. In his 2007 Prize Lecture (2007), which totaled ten pages, Hurwicz devotes only two paragraphs to discussing his pre-1972 work on mechanism design.
31. Hurwicz 1991, p. 83. In a personal and family history written by Hurwicz, dated March 28, 1999, he provides his dates in Geneva as 1939–40 (Leonid Hurwicz Papers, Perkins Library, Duke University).
32. Hurwicz 1955, p. 3.
33. Clearly, the most celebrated case was the U.S. FCC spectrum auctions, beginning in 1994, but this case requires careful consideration of the complex issues raised by it, so we will revisit it in chapter 15.
34. Grether et al. 1989. A much-condensed version of the report was published years earlier in the non-peer-reviewed *AER Papers and Proceedings* (Grether et al. 1981).
35. Levine was a Law and Economics Fellow at the University of Chicago Law School, and later served on the faculty at University of Southern California, which was a center of Chicago-style law and economics. Importantly, his position carried a joint appointment to Caltech. Plott has recounted the immediate circumstance leading to his work with Levine:

> Knowing of my interest in designing a "good" decision process, a colleague, Mike Levine, asked for help on a very practical problem, designing the agenda for a flying club which was preparing to decide on the fleet of airplanes that it would operate. Mike was a member of the club and was in charge of the agenda that was to be used at the meeting. By explaining the many impossibility theorems to Mike, I convinced him that there was no uniquely best agenda. Instead, there were many good agenda and each could lead to a different outcome. The thing to do was to decide which of the "good" agenda would lead to a choice he liked best. I designed an agenda using tools and intuition from previous experiments and Mike used it in the meeting of the flying club. The result was a success in the sense that the group chose the option that Mike wanted." (Plott, 2001, pp. xiv–xv)

36. See Plott 2005, p. 201; Plott 2014, p. 349; Svorenčík and Maas 2016, p. 99.
37. See Lee 2015; Plott 2014.

38. "The development of experiments in the early 1970s was driven by curiosity about the power of institutions to shape collective choice, much of which was stimulated by the work of James Buchanan and Gordon Tullock together with the broad issues of pubic choice and political science" (Plott 2014, p. 351). Plott himself later joined the MPS.
39. The Public Choice Society became an important staging ground for this program. See Svorenčík and Maas 2016, pp. 49–52.
40. See Smith 2008a, pp. 276–279.
41. Plott 1979, p. 139.
42. See McKee 1990. A faithful adherence to Vickrey's (1961) logic would suggest charging a price equal to the highest losing bid, not the lowest winning bid. We discuss Vickrey's work on auctions in more detail in chapter 12.
43. For example, they could meet the goal of servicing small communities by prioritizing airlines who agreed to serve routes to them. For an argument that even with this larger roster of goals, the Polinomics report failed to address some of the most pressing concerns at high traffic airports (e.g., the need to limit the noise generated by increased reliance on large jet aircraft). See Bailey et al. 1986, pp. 182–183.
44. Rassenti et al. 1982; see also Rassenti 1982.
45. We discuss the nature of this information processing problem in more detail in chapter 13.

Chapter 10

1. North 1977, p. 710.
2. Coase 1988, p. 7.
3. Arrow and Hahn 1971, p. 348.
4. Rosenbaum 2000.
5. A symptom of the general oblivion to market structures is the urban myth about the early neoclassical theory of Walras being inspired by the Paris Bourse. A good historian such as Walker (2001) makes short work of this fairy tale. On Edgeworth, see Mirowski 1994.
6. See Edwards et al. 2011, p. 1407; Mirowski 2002.
7. See Krippner 2011; Quinn 2010; Pardo-Guerra and MacKenzie 2014; Pardo-Guerra 2010.
8. Callon, quoted in Barry and Slater 2002, p. 300.
9. Mirowski and Nik-Khah 2007, 2008.
10. Boldyrev and Ushakov 2016.
11. Levine and Lippman 1995, p. xiv.

12. Also included under this heading is "Perfect Competition," "Monopoly," and "Oligopoly." Some, like Paul Klemperer (2004), have stressed a close relationship between studies of auctions and studies of perfect competition, monopoly, and oligopoly; texts in industrial organization often include chapters on the topic. See also Laffont and Tirole 1993. At the University of Chicago, market design is understood to be an extension of price theory. See http://home.uchicago.edu/weyl/PTMD_syllabus.pdf.
13. See www.aeaweb.org/jel/guide/jel.php.
14. "Market designers typically do not try to design a market all of whose equilibria accomplish something, but rather try to design a marketplace with a good equilibrium, and then try to achieve that equilibrium" (Vulkan et al. 2012, p. 3).
15. Levine and Lippman 1995, p. xxvii.
16. Börgers 2015. He continues: "Game theory takes the rules of the game as given, and it makes predictions about the behavior of strategic players. The theory of mechanism design is about the optimal choice of the rules of the game" (p. 2).
17. Milgrom 2004, p. 21.
18. See esp. Krishna 2002, pp. 61–82.
19. See http://theoryclass.wordpress.com/2014/01/22/market-design-class-lecture-1/. Even "game theory" assumes an unfamiliar sense at the hands of Roth, as he now denies the usefulness of distinguishing between cooperative and noncooperative game theory (Vulkan et al. 2012, p. 2).
20. While this does not mean that one can never find a present-day example of someone taking inspiration from the Walrasian program (e.g., Mount and Reiter 2002), it is less significant for present-day market design than the other programs.

Chapter 11

1. See www.nobelprize.org/nobel_prizes/economics/laureates/2007/press.html.
2. Hurwicz and Reiter 2006, p. 250.
3. Ibid., p. 26. The fantasy continues in the next sentence after this quote: "Less formal communications, such as business letters or memos, can be represented in the same way, if we abstract from chit-chat."
4. This is revealed in the Hurwicz interview in Feiwel 1987, p. 262: "Can one in some sense 'design' the economic system to have a more universal property of stability? . . . [M]uch of this work goes in the direction of designing a convergent computational system rather than designing a mechanism that could be applied in a real economy."

5. "Indeed, a very loose interpretation would say that this result goes beyond Hayek's claim: it asserts that, in a specified sense (dimension of the message space), you cannot do better than with the perfectly competitive mechanism" (Leonid Hurwicz, Letter to Fritz Machlup, March 12, 1982, Leonid Hurwicz Papers, Perkins Library Archives, Duke University, Box 20.
6. Reiter 1977, p. 277.
7. Ibid., p. 230. In some places, practitioners of Walrasian mechanism design were brought to identify the minimization of informational costs as the key economic problem of information (Hurwicz 1960; Hurwicz 1969, p. 174; Hurwicz 1986, p. 250; Reiter 1977.
8. Hurwicz 1977; Mount and Reiter 1974.
9. Strictly speaking, this is a bit too simple of a characterization owing to the possibility of "smuggling" a great deal of information into one real number (Hurwicz 1969, p. 515). Mount and Reiter dispensed with this problem by imposing further smoothness restrictions on the set of admissible messages. They introduced a measure of the size of a topological space (whereas commodity space is Euclidean) for the purpose of supporting a wider variety of messages (Mount and Reiter 1974, pp. 166–167).
10. Sometimes, the Walrasian mechanism designers have related communications issues to the "real economy" by arguing "fewer resources will be required to operate the system when the dimension of the message space (i.e., the number of message variables) is smaller" (Hurwicz 1986, p. 250), a condition owing to an assumed costliness of channel capacity or increased difficulty in calculation (Hurwicz 1960; Mount and Reiter 1974).
11. See Hurwicz and Reiter 2006, pp. 27–28.
12. Fritz Machlup, Letter to Leonid Hurwicz, April 5, 1982, Leonid Hurwicz Papers, Perkins Library Archives, Duke University, Box 20.
13. Mount and Reiter 2002.

Chapter 12

1. Vickrey 1960, pp. 517–519.
2. According to Roger Myerson (2004, p. 1818), Vickrey had produced "the one great paper with a truly modern treatment of information before Harsanyi."
3. Quoted in Dreze 1998.
4. One example of this shared enthusiasm is Wilson's contribution to the study of "decentralization under uncertainty" (Wilson 1969b). For his contributions to social choice theory, see Wilson 1969a, 1969c.
5. On Wilson as a student of Raiffa, see Raiffa 2002.
6. See Khurana 2010; Augier and March 2011; Nik-Khah 2011; Fourcade and Khurana 2013. According to Raiffa, at the time of his recruitment to Harvard

NOTES

from Columbia, "I really didn't know anything about business and the only reason I decided to go to Harvard was because of the Statistics Department. They were willing to double my Columbia salary" (Feinberg 2008, p. 142).
7. McGrayne 2011, p. 146.
8. See Feinberg 2008.
9. Raiffa 2002, p. 2; Feinberg 2008, p. 147.
10. Wilson 1996, p. 297.
11. Holmstrom et al. 2002, p. 4.
12. Interestingly, Alvin Roth would also study under Wilson.
13. Holmstrom et al. 2002.
14. Rothkopf 1969, p. 362.
15. Milgrom and Weber 1982 is generally regarded as providing the canonical model of the Bayes-Nash approach (Krishna 2002, pp. 83–102; McMillan 1994, p. 146). The following model is taken from Milgrom 2004, esp. pp. 45–46, 195–198. Here, we omit the post hoc attempts to render work in this tradition in language similar to that of the Walrasian School of Design.
16. Because the sum of prices paid equals zero, the price term drops out of the aggregate value function.
17. Because the auction game is assumed to be symmetric, it is possible to identify an equilibrium strategy profile by focusing entirely on the strategy of a single bidder.
18. That this is not the *only* possible equilibrium bidding strategy would certainly complicate the task required of even the most sophisticated bidders, dutifully committed to playing their roles in the symmetric game, perhaps fatally. In what follows, we elect to ignore the well-known difficulties presented by multiple equilibria to focus on problems specific to the informational setup of the Bayes-Nash program.
19. Here we use the term "steps" so as to avoid confusion with the term "stages," which assumes a specific technical meaning within the game theoretic literature that does not match ours here.
20. I.e., $t^{(n)} = s, \forall_{n=1,\ldots,N}$. This is why all the values in the valuation function are equal to s.
21. In assuming this, the bidder would of course be "wrong" if all bidders did in fact immediately drop before he did, but in terms of the effect on his received payoff it is of no matter, since he would never acquire the item at a price greater than his reservation value. Employing the assumption amounts to inferring only that all remaining bidders' types are *at least* high enough to rationally remain in the auction to this point. Hence, the strategy adopted would be "right" in that it maximizes the bidder's payoff within the context of the symmetric game (meaning that all other participants follow the same strategy, and act on information released within the auction in the same way).

NOTES

22. A fundamental assumption enabling this value emendation is "affiliation" (roughly, that bidder "types" are related in such a way that a higher-valued type observed in another bidder makes it more likely that one's own valuation is higher valued).
23. See Capen et al. 1971. Rothkopf 2000 is especially clear on this early work. Robert Aumann reports: "Wilson has consulted for oil companies bidding on off-shore oil tracts worth upwards of 100 million dollars each" (Van Damme, 1998, p. 183).
24. Ashenfelter 1989.
25. Mikoucheva and Sonin 2004, p. 278. See also Maskin 1992; Jehiel and Moldovanu 2001; Jackson 1999, 2009.
26. Paul Milgrom interview, in Bowmaker 2012, p. 338.

Chapter 13

1. Plott 1994, p. 3.
2. Smith 2008a, p. 289.
3. Smith 2006, p. xi.
4. One of us has discussed this development in a different context as a shift of economics into the realm of the "cyborg sciences" (see Mirowski 2002).
5. During the period it was the epicenter of studies in experimental economics, George Mason University established a graduate certification in economic system design; today, Chapman University (home to Vernon Smith) offers an MS in economic system design.
6. The following is adopted from De Vries and Vohra 2003.
7. Mechanism design has matured over the past 20 years by focusing on incentive compatibility and political viability. The analysis has usually been carried out under the working assumption that infinite computing capacity is always available. Any computation required of the individuals or of the system can be instantaneously and correctly completed. Of course, any expert in organizational computing knows this is clearly wrong. (Ledyard 1993, p. 122)

   Ledyard devoted the rest of his article to suggesting how to bring together research in organizational computing and the economics of market design.
8. Smith 2006, p. xi.
9. Porter et al. 2003, p. 11154.
10. Bykowsky et al. 2000, p. 218; Porter et al. 2003, pp. 11154–11155.
11. The technological argot has increasingly encouraged participants to describe their markets in terms usually reserved for machines. The drive to patent markets is inducing game theorists to also begin to talk like this. We discuss the efforts of game theorists to patent their auctions below. This technological language also permeates the "performativity" theorists in the social

NOTES

studies of science (MacKenzie et al. 2007). This is yet another way sociologists of science follow the moves of economists, rather than offer alternative understandings of their work.

12. Banks et al. 1989, pp. 2–3; Ledyard et al. 1997, p. 656.
13. Plott 2001, p. xvi.
14. Roth 2015, esp. pp. 134–144.
15. Smith 2008b, pp. 121–122.
16. Plott 1994, p. 4.
17. McCabe et al. 1999, p. 810.
18. Roth 2002a, p. 1372.
19. See Mirowski 2007, p. 219.

Chapter 14

1. In his latest book, Alvin Roth provides another such acknowledgment:

    In chapter 1, I made the analogy between a free market with effective rules and a wheel that can rotate freely because it has an axle and well-oiled bearings. I could have been paraphrasing the iconic free-market economist Friedrich Hayek. . . . He understood that markets need effective rules to work freely. . . . Hayek also understood that there is a place for economists to help in understanding how to design markets (Roth 2015, p. 226).

2. Maskin 2015, p. 251.
3. For initial responses to the award, see Peter Boettke, "A Market Nobel," *Wall Street Journal*, October 16, 2007, at www.wsj.com/articles/SB119249811353060179; Alex Tarrabok, "What Is Mechanism Design?," *Reason.com*, at https://reason.com/archives/2007/10/16/what-is-mechanism-design). For one example of a paper devoted to establishing Hayek's influence on the neoclassical orthodoxy, see Skarbek 2009.
4. Regarding the former, see Boettke 2002. Regarding the latter, see Lavoie 1986.
5. Boettke and O'Donnell 2013.
6. Here and there one encounters the claim that game theory should be viewed as an outgrowth of a broader Austrian tradition, yet such claims are advanced halfheartedly and without reference to Hayek's work. See Foss 2000; Kiesling 2015.
7. Boettke and Coyne 2015. The roster of invitees to the George Mason conference included Israel Kirzner, Edmund Phelps, Vernon Smith, and Maskin—which gives some indication of the insights that can be expected from this project.
8. Mount and Reiter 1974, p. 163.

NOTES

9. Wilson 1996, p. 296.
10. McAfee and McMillan 1987, pp. 721–722.
11. Smith 1991, p. 811.
12. Smith 2010, p. 5n7.
13. Smith 2015, p. 242.
14. Smith 2006, p. xii.
15. As we have seen, they failed to convince many Austrians of the same, at least when it came to the activities of the Walrasian and Bayes-Nash Schools. Their reactions to the design activities of the Experimentalists have been more muted, possibly because of formal affiliation with the neoliberal project: both Vernon Smith and Charles Plott are members of the Mont Pèlerin Society.
16. Panning out, one might well include the MPS here; certainly it functioned to keep Hayek's concerns at the forefront of the Experimentalists' minds.
17. Matthew Jackson, "Background on the NSF/CEME Decentralization Conference Series," at http://web.stanford.edu/~jacksonm/history.htm.
18. See www.lsa.umich.edu/cscs/events/annualevents/decentralizationconference/pastconferences. A full roster of participants in that first meeting gives a sense of how closely the effort was to the design literature: Masahiko Aoki, Masanao Aoki, Jerry Green, Theodore Groves, Terry Hogan, Leonid Hurwicz, Mordecai Kurz, John Ledyard, Jacob Marschak, Stanley Reiter, Roman Weil; attendees from University of California, Berkeley (where it was held), included Roy Radner, John Harsanyi, Thomas Marschak, David Gale, and C. B. McGuire. All our information about participation at the Decentralization Conference Series is taken from conference programs posted to the above website.
19. This may have been due in part to his close proximity to John Harsanyi, who was then employed in the School of Management at UC Berkeley.
20. For example, one session at the 1986 meetings was devoted to "Connections between the economic theory of decentralized resource allocation and the theory of parallel/distributed computing," and at the 1993 meetings a joint session was held with the Conference on Coordination and Complexity at UC Berkeley.
21. "My impression is that the word decentralization [in the title of the conference] reflects the fact that the starting point in many of the problems addressed by the series is that the necessary information starts in a decentralized state. . . . I think that perhaps these systems were viewed as alternatives to centralized or planned economies when the conference series was first funded, during the cold war" (Matthew Jackson, "Background on the NSF/CEME Decentralization Conference Series," at http://web.stanford.edu/~jacksonm/history.htm).
22. McCabe et al. 1991.

23. Maskin 1992. On this interpretation of Maskin's result, see, e.g., Goeree and Offerman 2003.

Chapter 15

1. McMillan 2004, pp. 73–74. See also McMillan 2003, from which the phrase "free lunch redistribution" is taken.
2. Michel Callon 2007 has looked to the FCC auctions for support of his "performativity thesis," discussed in chapter 10. The performativity branch of science and technology studies is described in MacKenzie et al. 2007. For a critique, see Mirowski and Nik-Khah 2007.
3. Prior to the auctions, the FCC relied on comparative hearings and lotteries to assign spectrum licenses.
4. FCC 1993, paragraph 34; FCC 1994, paragraph 70.
5. The FCC's Office of Plans and Policy was handed the task of drafting recommendations for the auction. Accounts from this perspective are provided by Kwerel and Rosston 2000; Kwerel 2004.
6. One plan for the auction of licenses called for a sequence of English auctions (Weber 1993a, 1993b), a second called for a sequence of Japanese auctions (Nalebuff and Bulow 1993a, 1993b), and a third called for simultaneous sales of licenses (McAfee 1993a, 1993b; Milgrom and Wilson 1993a, 1993b). An English auction is one for which prices increase, with the bidder placing the highest bid winning the item. A Japanese auction is similar to an English auction, but all participants are considered active bidders until they drop out. Some proposals insisted on admitting bids for bundles of geographically linked licenses, whereas others favored restricting bids to individual licenses only.
7. We first discussed this problem, inherent to the Bayes-Nash School, in chapter 12. See also Banks et al. 2003; Goeree and Offerman 2003.
8. McAfee and McMillan 1996, p. 171; McMillan et al. 1997, p. 429.
9. Bykowsky and Cull 1994.
10. It will become apparent below that some game theorists supported package bidding.
11. McAfee 1993a, pp. 12–14; Milgrom and Wilson 1993a, pp. 8–13; Milgrom and Wilson 1993b, pp. 4–5.
12. Plott 1997, p. 606.
13. Ausubel and Milgrom 2006, pp. 79–80; emphasis added.
14. The FCC eventually enlisted the services of John McMillan, who produced a report for the FCC that was published in revised form as McMillan 1994. For a discussion of the controversial aspects of this report, see (Nik-Khah 2005).
15. This case is made in much greater detail in Nik-Khah 2005.

16. Murray 2002, pp. 274–275.
17. The original plan called for allocating licenses in three auctions, to be conducted over a two-year period. The FCC was eventually forced to conduct eleven auctions over a ten-year period. The process of re-auctioning finally concluded in February 2005—a full decade after the auctions commenced.
18. The success of large telecoms in circumventing the FCC's "designated entity" provisions, most notoriously by establishing shell companies (Cramton et al. 2002; Labaton and Romero 2001), goes some way to explain why certain large telecommunications companies would voluntarily extol the FCC's highest-valued-user criterion, so long as it was interpreted as "willingness and ability to pay the most"—and they were not required to pay as much as they were willing and able (Mirowski and Nik-Khah 2007).
19. Copps 2004; Meister 1999, pp. 76–77.
20. Murray 2002, pp. 289–291.
21. Labaton and Romero 2001.
22. Thelen 1995.
23. Ledyard et al. 1997.
24. Helm 1994.
25. CNN Business Morning 1994.
26. Milgrom 2004, p. 23.
27. Thelen 1995.
28. McMillan, 1995, p. 194).
29. Klemperer 2004, p. 170.
30. McMillan 2002, p. 14.
31. Cramton 2002, p. 3.
32. Market Design Incorporated (www.market-design.com) "offers consulting services in the design of auction markets." Spectrum Exchange (www.spectrum-exchange.com) boasts it is "creating value through the efficient exchange of spectrum." Other companies include Power Auctions LLC—"No other organization can match the depth and breadth of our experience and success in high-stakes auctions" (www.powerauctions.com/company), Efficient Auctions LLC—"a provider of intellectual property for advanced auction applications" (http://efficientauctions.com/index.htm), and Auctionomics—"a high stakes auction consulting and software firm offering simple, innovative and economically sound solutions to complicated problems" (www.auctionomics.com/).
33. Roth 2002b. Or, as Eli Noam puts it, "The FCC auctions have . . . benefited from the contributions of game theorists grateful for a field of recognition after the end of the Cold War" (Noam 1995, p. 2).
34. Available at the Auctionomics website (www.auctionomics.com/). Multiple market design companies got into the act: "Power Auctions and its associate,

Efficient Auctions LLC, have an active program of developing and obtaining intellectual property rights for new auction technology.... They have published international patent applications related to dynamic auctions, and have other patents pending. The PowerAuctions software system utilizes many aspects of these patented technologies, and other aspects are currently under development." Available at the Power Auctions website (www.power-auctions.com/ip).

35. With what one imagines to be equal parts disapproval and amusement, Vernon Smith observed:

The Federal Communication Commission's Simultaneous Multiple Round auction evolved over a sequence of field applications in which weaknesses and defects revealed in each application led to "fine tuning," followed by the observation of further problem leading to new "fixes," and so on. Each "fix," designed to limit a particular strategic exploitation, tended also to generate complexity and its attendant higher transactions' cost. This was precisely what had been learned [in 1988] in the laboratory in a series of elementary experiments. (Smith 2006, p. xiii).

Chapter 16

1. See, however, Mirowski 2013.
2. In the present chapter we do not discuss the circumstances in Europe. Future research is needed to determine how market designers influenced the response to the crisis there.
3. Swagel 2009; Sorkin 2009.
4. "Break the Glass Bank Recapitalization Plan," April 15, 2008, at www.scribd.com/doc/21266810/Too-Big-To-Fail-Confidential-Break-the-Glass-Plan-from-Treasury.
5. "Secretary Paulson's intent to use TARP to purchase assets reflected a philosophical concern with having the government buy equity stakes in banks: he saw it as fundamentally a bad idea to have the government involved in bank ownership" (Swagel 2009, p. 50).
6. Oliver Armantier and James Vickery of the New York Fed delivered the baseline auction proposal on September 20; during the following week, the Treasury and the New York and Washington Feds reached out to the academic market designers Lawrence Ausubel, Peter Cramton, Jacob Goeree, Charles Holt, Paul Milgrom, Jeremy Bulow, and Jonathan Levin. See Armantier et al. 2011; Klemperer 2010.
7. Ferguson and Johnson 2009, pp. 28–29. See "Bernanke's Comments on Asset Auction Process," September 23, 2008, at www.reuters.com/article/2008/

09/23/financial-bailout-bernanke-auctions-idUSN2338396920080923. The concern was with "mark to market" accounting rules, under which low prices might make banks appear insolvent.

8. For example: "Treasury is talking with the experts you would expect—prominent academics who have designed auctions.... Treasury is committed to get the market price as best it can." Swagel, quoted on Greg Mankiw's blog, September 25, 2008, at http://gregmankiw.blogspot.com/2008/09/defense-of-paulson-plan.html. Whereas the quote is unattributed on this blog entry, Swagel has subsequently made clear that he was its author (Swagel 2009, p. 47).

9. For an example of the latter, see Tim Ryan, "Lesson from Saving and Loan Rescue," *Financial Times*, September 24, 2008, at www.ft.com/content/8e19c058-8a35-11dd-a76a-0000779fd18c. Tim Ryan was the president and CEO of the Securities Industry and Financial Markets Association, a lobbying group.

10. Ausubel and Cramton 2008a. Cramton made clear that he shared Bernanke's concern in an NPR interview with David Kestenbaum: "If the price [for a toxic asset] was too low then the banks would collapse and we would still have a mess." See "Complicated Reverse Auction May Aid In Bailout," October 10, 2008, at www.npr.org/templates/story/story.php?storyId=95591129.

11. Swagel 2009, p. 56.

12. The fact they were *academic* economists was significant. Swagel noted that Wall Street economists were also in favor of the TARP, but acknowledged that people would be suspicious of their judgments. (Swagel, quoted on Mankiw's blog, September 25, 2008, at http://gregmankiw.blogspot.com/2008/09/defense-of-paulson-plan.html.

13. Ausubel and Cramton 2008a, p. 2.

14. Ausubel and Cramton 2008b; Klemperer 2010; Armantier et al. 2011.

15. Swagel 2009; Armantier et al. 2011.

16. Ausubel and Cramton 2008b, p. 10.

17. And, indeed, the studies that Ausubel and Cramton draw upon to get their 97% figure (Kagel and Levin 2001, 2009) provided experimental treatments of *private value* auctions.

18. While the criterion does make sense in the case of private value auctions, one can never suffer from the winner's curse in such cases, again *by definition*, and therefore the argument to prefer the clock auction on the grounds of information aggregation is nonsense. Since the market designers' claim that one could avert the crisis by increasing information about the value of assets implied that the assets must be common valued (or else the link between auction performance and crisis aversion is severed), the "efficiency" evidence is especially misleading.

NOTES

19. Matthew Philips, "Gaming the Financial System," *Newsweek*, November 18, 2008, at www.thedailybeast.com/newsweek/2008/11/17/gaming-the-financial-system.html.
20. Lawrence Ausubel and Peter Cramton, "Auction Design for the Rescue Plan," presentation, October 5, 2008, at www.cramton.umd.edu/papers2005-2009/ausubel-cramton-auction-for-rescue-plan-slides.pdf.
21. Paulson 2010, pp. 258, 264.
22. Ibid., p. 334; see also Swagel 2009, p. 50.
23. Paulson 2010, pp. 323–326.
24. Ibid., pp. 363–368; Swagel 2009, pp. 50–52.
25. Paulson 2010, p. 389; Swagel 2009, p. 58.
26. Ausubel and Cramton 2008c.
27. Davies and McGoey 2012, p. 77.

Chapter 17

1. He came in only four spots below Steven Levitt's *Freakonomics* blog. See http://www.onalyticaindexes.com/2013/07/31/top-200-influential-economics-blogs-aug-2013/. Point your browser at: http://noahpinionblog.blogspot.com/. The last time we looked (September 19, 2015), he was still defending the intellectual legitimacy of the efficient markets hypothesis. Is this the irony David Foster Wallace warned us about?
2. He is willing to grant the point, but only for macroeconomists: "So if you really feel you must get out your rake or pitchfork and storm the gates of the economists who fiddled while our economy burned, go ahead. Just make sure that the people's whose heads you are calling for are not in that vast silent majority who are working diligently on the small but solvable problems of 'microeconomics.' The People at whom you are angry are called 'macroeconomists.'" See http://theweek.com/article/index/255013/why-economists-get-a-bad-rap.
3. See http://qz.com/208402/economics-can-do-many-things-but-it-cannot-help-the-economy/.
4. One might protest that his willingness to throw macroeconomics under the bus is pretty distinctive, but we believe that a willingness to grant such arguments against macro, while maintaining all is well with micro, has been actually pretty common since the crisis. Hence, yet another reason why we opted to focus on market design.
5. Deringer 2015, pp. 655–656.
6. Poets like Auden would cast the aspersions right back at the economists.
7. Revealingly, the lone example cited by Deringer as evidence of this supposed borrowing (Hong et al. 2007) merely name-checks Thomas Kuhn, only to

fall back on a standard-grade model of the Bayes-Nash variety. We discussed the Bayes-Nash theoretical tradition in chapter 12.
8. That is, in the first instance following the transition from mechanism design to market design.
9. Smith 2010, p. 5n7.
10. This raises the question as to whether the Austrians accept the orthodoxy's design turn. From time to time, one will encounter some grumbling, usually from among the ranks of the self-identified Austrians and fellow travelers. Doesn't engineering markets amount to "social engineering?" The reaction is not shared by all Austrians; among individual Austrians, reactions to market design vary over time and circumstance. But at least one problem bedevils any attempt to critique market design from an Austrian standpoint. Notwithstanding all the fuss about "spontaneous order," Austrians acknowledge that The Market is a constructed entity, requiring a framework within which to flourish. If Austrians persisted in philosophical reflection, they might come to a fuller appreciation of the significance of their revulsion toward market design.
11. McMillan 2002, p. 6.
12. Noah Smith, "Economists used to be the priests of free markets—now they're just a bunch of engineers." Quartz. At http://qz.com/208402/economics-can-do-many-things-but-it-cannot-help-the-economy/.
13. Tkacik 2010.

BIBLIOGRAPHY

Agar, John. 2012. *Science in the 20th Century and Beyond*. Cambridge: Polity.
Akerlof, George. 1970. "The Market for Lemons." *Quarterly Journal of Economics* 84: 488–500.
Akerlof, George. 2002. "Behavioral Macroeconomics and Microeconomic Behavior," at: http://www.nobelprize.org
Ambler, Eric. 2002. *Epitaph for a Spy*. New York: Vintage.
Ames, Edward. 1981. "On Forgetting Economics with Em Weiler." In George Horwich and James Quirk, eds., *Essays in Contemporary Fields of Economics*, pp. 355–360. West Layfayette, IN: Purdue University Press.
Angner, Erik. 2012. *A Course in Behavioral Economics*. London: Palgrave.
Arena, Richard, Agnes Festre, and I. Lazaric, eds. 2012. *Handbook of Knowledge and Economics*. Northampton, MA: Edward Elgar.
Armantier, Olivier, Charles Holt, and Charles Plott. 2011. "A Procurement Auction for Toxic Assets with Asymmetric Information." CalTech Social Science Working Paper 1330R.
Arrow, Kenneth. 1951. *Social Choice and Individual Values*. New York: Wiley.
Arrow, Kenneth. 1962. "Economic Welfare and the Allocation of Resources for invention." In Richard Nelson, ed., *The Rate and Direction of Inventive Activity*, pp. 609–626. Princeton, NJ: Princeton University Press.
Arrow, Kenneth. 1984. *Collected Papers of Kenneth J. Arrow. Vol. 4: The Economics of Information*. Cambridge, MA: Harvard University Press.
Arrow, Kenneth. 1996. "The Economics of Information: an exposition." *Empirica* 23: 119–128.
Arrow, Kenneth. 2009. "Some Developments in Economic Theory since 1940: An Eyewitness Account." *Annual Review of Economics* 1: 1–16.

BIBLIOGRAPHY

Arrow, Kenneth, David Blackwell, and Meyer Girshick. 1949. "Bayes and Minimax Solutions of Sequential Decision Problems." *Econometrica* 17: 213–244.

Arrow, Kenneth, and Frank Hahn. 1971. *General Competitive Analysis*. New York: North-Holland.

Ashenfelter, Orley. 1989. "How Auctions Work for Wine and Art." *Journal of Economic Perspectives* 3(3): 23–36.

Aspray, William. 1985. "Scientific Conceptualization of Information: A Survey." *Annals of the History of Computing* 7(2): 117–140.

Augier, Mie, and James March. 2011. *The Roots, Rituals and Rhetorics of Change*. Stanford, CA: Stanford University Press.

Aumann, Robert. 1976. "Agreeing to Disagree." *Annals of Statistics* 4: 1236–1239.

Ausubel, Lawrence, and Peter Cramton. 2008a. "Auction Design Critical for Rescue Plan." *Economists' Voice* 5(5). At: http://works.bepress.com/cramton/3/.

Ausubel, Lawrence, and Peter Cramton. 2008b. "A Troubled Asset Reverse Auction." Working Paper, University of Maryland. At http://works.bepress.com/cramton/9/.

Ausubel, Lawrence, and Peter Cramton. 2008c. "Auctions for Injecting Bank Capital." Working Paper, University of Maryland. At http://works.bepress.com/cramton/8.

Ausubel, Lawrence, and Paul Milgrom. 2006. "Ascending Proxy Auctions." In P. Cramton, Y. Shoham, and R. Steinberg, eds., *Combinatorial Auctions*, pp. 79–98. Cambridge, MA: MIT Press.

Babe, Robert, ed. 1994. *Information and Communication in Economics*. Boston: Kluwer.

Bailey, Elizabeth, David Graham, and Daniel Kaplan. 1986. *Deregulating the Airlines*. Cambridge, MA: MIT Press.

Banks, Jeffrey, John Ledyard, and David Porter. 1989. "Allocating Uncertain and Unresponsive Resources: An Experimental Approach." *RAND Journal of Economics* 20(1): 1–25.

Banks, Jeffrey, Mark Olson, David Porter, Stephen Rassenti, and V. Smith. 2003. "Theory, Experiment, and the Federal Communications Commission Spectrum Auctions." *Journal of Economic Behavior and Organization* 51(3): 303–350.

Barry, Andrew, and Don Slater. 2002. "Technology, Politics and the Market: An Interview with Michel Callon." *Economy and Society* 31(2): 285–306.

Basili, Marcello, and Carlo Zappia. 2009. "Shackle and Modern Decision Theory." *Metroeconomica* 60: 245–282.

Becker, Gary. 1964. *Human Capital*. New York: NBER.

Becker, Gary. 2002. "The Age of Human Capital." In E. P. Lazear, ed., *Education in the Twenty-First Century*, pp. 3–8. Palo Alto, CA: Hoover Institution Press.

Bessner, Daniel, and Nicolas Guilhot. 2016. "A Decisionist History of the 20th Century." Unpublished manuscript.

Blackwell, David. 1953. "Equivalent Comparisons of Experiments." *Annals of Mathematical Statistics* 24: 265–272.

Blackwell, David, and Meyer Girshick. 1954. *Theory of Games and Statistical Decisions*. New York: John Wiley.

Blaug, Mark. 1996. *Economic Theory in Retrospect*, 5th ed. New York: Cambridge University Press.

Boettke, Peter. 2002. "Information and Knowledge: Austrian Economics in Search of Its Uniqueness." *Review of Austrian Economics* 15(4): 263–274.

Boettke, Peter, and Christopher Coyne. 2015. "Hayek's Nobel after 40 Years." *Review of Austrian Economics* 28(3): 221–223.

Boettke, Peter, and Kyle O'Donnell. 2013. "The Failed Appropriation of F.A. Hayek by Formalist Economics." *Critical Review* 25(3/4): 305–341.

Boldyrev, Ivan, and Alexy Ushkanov. 2016. "Adjusting the Model to Adjust the World: Constructive Mechanisms in Postwar General Equilibrium Theory." *Journal of Economic Methodology* 23(1): 38–56.

Börgers, Tilman. 2015. *An Introduction to the Theory of Mechanism Design*. New York: Oxford University Press.

Boulding, Kenneth. 1966. "The Economics of Knowledge and the Knowledge of Economics." *American Economic Review* 56: 1–13.

Boulding, Kenneth. 1968. *Beyond Economics: Essays on Society, Religion and Ethics*. Ann Arbor: University of Michigan Press.

Bowmaker, Simon. 2012. *The Art and Practice of Economic Research*. Northampton, MA: Edward Elgar.

Boyle, James. 2000. "Cruel, Mean or Lavish?" *Vanderbilt Law Review* 53: 2007–2039.

Burgin, Angus. 2012. *The Great Persuasion*. Cambridge, MA: Harvard University Press.

Butos, William. 2010. *The Social Science of Hayek's Sensory Order*. Bingley, UK: Emerald Group.

Bykowsky, Mark, and Robert Cull. 1994. "Personal Communications Services Auction: Further Analysis." NTIA Office of Policy Analysis and Development Staff Paper, PP Docket No. 93-253.

Bykowsky, Mark, Robert Cull, and John Ledyard. 2000. "Mutually Destructive Bidding: The FCC Auction Design Problem." *Journal of Regulatory Economics* 17(3): 205–228.

Caldwell, Bruce. 2004. *Hayek's Challenge*. Chicago: University of Chicago Press.

Callon, Michel. 2007. "What Does it Mean to Say that Economics Is Performative?" In D. MacKenzie, F. Muniesa, and L. Siu, eds., *Do Economists Make Markets?*, pp. 311–357. Princeton, NJ: Princeton University Press.

Capen, E. C., R. V. Clapp, and W. M. Campbell. 1971. "Competitive Bidding in High Risk Situations." *Journal of Petroleum Technology* 23(6): 641–653.

Carnap, Rudolf. 1947. *Meaning and Necessity*. Chicago: University of Chicago Press.

Carrick-Hagenbarth, Jessica, and Gerald Epstein. 2012. "Dangerous Interconnectedness: Conflicts of Interest, Ideology and the Financial Crisis." *Cambridge Journal of Economics* 36: 43–63.

Cartwright, Edward. 2011. *Behavioral Economics*. New York: Routledge.

Christ, Carl. 1994. "The Cowles Commission's Contribution to Econometrics at Chicago." *Journal of Economic Literature* 32: 30–59.

CNN Business Morning. 1994. *RTV Reports*, December 6.

Coase, Ronald. 1974. "The Market for Goods and the Market for Ideas." *American Economic Review, Papers and Proceedings*, May: 384–391.

Coase, Ronald. 1988. *The Firm, the Market and the Law*. Chicago: University of Chicago Press.

Colander, David, Ric Holt, and J. B. Rosser, eds. 2004. *The Changing Face of Economics*. Ann Arbor: University of Michigan Press.

Conlisk, John. 1996. "Why Bounded Rationality?" *Journal of Economic Literature* 34: 669–700.

Copps, Michael. 2004. "Statement of Commissioner Michael J. Copps." In Federal Communications Commission (FCC), Report and Order and Further Notice of Proposed Rulemaking, FCC Docket No. 04-166.

Cowan, R., P. David, and D. Foray. 2000. "The Explicit Economics of Knowledge: Codification and Tacitness." *Industrial and Corporate Change* 9: 211–253.

Cramton, Peter. 2002. "Introduction to Chapter." In Bengt Holmstrom, Paul Milgrom, and Alvin Roth. Berkeley, eds., *Game Theory in the Tradition of Bob Wilson*. Berkeley, CA: Bepress. At www.bepress.com/wilson/.

Cramton, Peter, Allan Ingraham, and Hal Singer. 2002. "The Impact of Incumbent Bidding in Set-Aside Auctions: An Analysis of Prices in the Closed and Open Segments of FCC Auction 35." Criterion Economics Working Paper No. 02-07.

Dang, Tri, Gary Gorton, and Bengt Holmstrom. 2012. "Ignorance, Debt and Financial Crises." Cowles Working Paper. At www.columbia.edu/~td2332/Paper_Ignorance.pdf.

Daston, Lorraine. 1978. "British Responses to Psycho-physiology, 1860–1900." *Isis* 69: 192–208.

Davies, Paul, and Niels Gregersen. 2011. *Information and the Nature of Reality*. New York: Cambridge University Press.

Davies, William, and Linsey McGoey. 2012. "Rationalities of Ignorance: on Financial Crisis and the Ambivalence of Neoliberal Epistemology." *Economy and Society* 41(1): 64–83.

Davis, John. 2006. "The Turn in Economics: Neoclassical Dominance to Mainstream Pluralism." *Journal of Institutional Economics* 2: 1–20.

Davis, John. 2008. "The Turn in Recent Economics and the Return of Orthodoxy." *Cambridge Journal of Economics* 32: 349–366.

Davis, John. 2011. *Individuals and Identity in Economics*. New York: Cambridge University Press.

De Vries, Sven, and Rakesh Vohra. 2003. "Combinatorial Auctions: A Survey." *INFORMS Journal on Computing* 15(3): 284–309.

DeGroot, Morris. 1986. "A Conversation with David Blackwell." *Statistical Science* 1: 40–53.

Dekel, Eddie, and Marciano Siniscalchi. 2014. *Epistemic Game theory*. At https://sites.google.com/site/eddiedekelsite/papers/game-theory.

Dekel, Eddie, Barton Lipman, and Aldo Rustichini. 1998. "Standard State Space Models Preclude Unawareness." *Econometrica* 66: 159–173.

DeMartino, George. 2011. *The Economists' Oath*. New York: Oxford University Press.

Deringer, William. 2015. "For What It's Worth: Historical Financial Bubbles and the Boundaries of Economic Rationality." *Isis* 106(3): 646–656.

Diamond, Arthur. 2012. "The Epistemology of Entrepreneurship." *Advances in Austrian Economics* 17: 111–142.

Diamond, Peter, and Michael Rothschild, eds. 1978. *Uncertainty in Economics: Readings and Exercises*. New York: Academic Press.

Dolfsma, Wilfred, and Luc Soete, eds. 2006. *Understanding the Dynamics of a Knowledge Economy*. Cheltenham: Edward Elgar.

Dorfman, Robert. 1960. "Operations Research." *American Economic Review* 50: 575–623.

Dreze, Jacques. 1998. "William S. Vickrey 1914–1996." *National Academy of Sciences Biographical Memoirs*. Washington, DC: National Academies Press. At www.nasonline.org/publications/biographical-memoirs/memoir-pdfs/vickrey-william.pdf.

Dumit, Joseph. 2012. *Drugs for Life*. Durham, NC: Duke University Press.

Edwards, Paul N., Lisa Gitelman, Gabrielle Hecht, Adrian Johns, Brian Larkin, and Neil Safier. 2011. "Historical Perspectives on the Circulation of Information." *American Historical Review* 116(5): 1393–1433.

Emmett, Ross. 1998. "What Is Truth in Capital Theory?" In John Davis, ed., *New Economics and its History*, supplement to Vol. 29: *History of Political Economy*, pp. 231–252. Durham, NC: Duke University Press.

Erickson, Paul. 2010. "Mathematical Models, Rational Choice, and the Search for Cold War Culture." *Isis* 101: 386–392.

Erickson, Paul, Judy Klein, Lorraine Daston, Rebecca Lemov, Thomas Sturm, and Michael Gordin. 2013. *How Reason Almost Lost Its Mind*. Chicago: University of Chicago Press.

Fagin, Ronald, Joseph Halpern, and Yoram Moses. 1995. *Reasoning about Knowledge.* Cambridge, MA: MIT Press.

Federal Communications Commission (FCC). 1993. Notice of Proposed Rulemaking (NPRM). FCC Docket No. 93-455.

Federal Communications Commission (FCC). 1994. Second Report and Order. FCC Docket No. 94-61.

Feinberg, Stephen. 2008. "The Early Statistical Years: 1947–1967 A Conversation with Howard Raiffa." *Statistical Science* 23(1): 136–149.

Feiwel, George, ed. 1987. *Arrow and the Ascent of Modern Economics.* New York: New York University Press.

Ferguson, Thomas, and Robert Johnson. 2009. "Too Big to Bail: The 'Paulson Put,' Presidential Politics, and the Global Financial Meltdown, Part II." *International Journal of Political Economy* 38(2): 5–45.

Folkers, Andreas. 2016. "Daring the Truth: Foucault, Parrhesia and the Genealogy of Critique." *Theory Culture and Society* 33: 3–28.

Foss, Nicolai. 2000. "Austrian Economics and Game Theory: A Stocktaking and an Evaluation." *Review of Austrian Economics* 13(1): 41–58.

Foucault, Michel. 2008. *The Birth of Biopolitics.* New York: Palgrave Macmillan.

Foucault, Michel. 2011. *The Courage of Truth.* New York: Palgrave Macmillan.

Fourcade, Marion, and Rakesh Khurana. 2013. "From Social Control to Financial Economics." *Theory and Society* 42: 121–159.

Fourcade, Marion, Etienne Ollion, and Yann Algan. 2015. "The Superiority of Economists." *Journal of Economic Perspectives* 29(1): 89–114.

Gentzkow, Matthew, and Jesse Shapiro. 2008. "Competition and Truth in the Market for News." *Journal of Economic Perspectives* 22: 133–154.

Geoghegan, Bernard. 2008. "The Historiographic Conception of Information: A Critical Survey." *IEEE Annals of the History of Computing* 30: 66–81.

Gerovitch, Slava. 2009. "The Cybernetics Scare and the Origins of the Internet." *Baltic Worlds* 2(1): 32–38.

Gigerenzer, Gerd. 2012. "What Can Economists Know?" At http://videos-with-links.info/watch/DdEEwoKkfMA/gerd-gigerenzer-what-can-economists-know-25.html.

Gigerenzer, Gerd, and David Murray. 1987. *Cognition as Intuitive Statistics.* Hillsdale, NJ: Lawrence Erlbaum.

Giocoli, Nicola. 2003. *Modeling Rational Agents.* Northampton, MA: Edward Elgar.

Giocoli, Nicola. 2009. "Three Alternative (?) Stories on the Late 20th Century Rise of Game Theory." *Studi e Note di Economia* 14(2): 187–210.

Gladwell, Malcolm. 2005. *Blink: The Power of Thinking without Thinking.* New York: Little Brown.

Gleick, James. 2010. "The Information Palace." At http://www.nybooks.com/blogs/nyrblog/2010/dec/08/information-palace/.
Gleick, James. 2011. *The Information*. London: Fourth Estate.
Gode, Dan, and Shyam Sunder. 1993. Allocative Efficiency of Markets with Zero-Intelligence Traders. *Journal of Political Economy* 101: 119–137.
Goeree, Jacob, and Theo Offerman. 2003. "Competitive Bidding in Auctions with Private and Common Values." *Economic Journal* 113(489): 598–613.
Goldstein, William, and Robin Hogarth. 1997. "Judgment and Decision Research: Some Historical Context." In William Goldstein and Robin Hogarth, eds. *Research on Judgment and Decision Making*, pp. 3–65. Cambridge: Cambridge University Press.
Grether, David, R. Mark Isaac, and Charles Plott. 1981. "The Allocation of Landing Rights by Unanimity Among Competitors." *American Economic Review Papers and Proceedings* 71(2): 166–171.
Grether, David, R. Mark Isaac, and Charles Plott. 1989. *The Allocation of Scarce Resources: Experimental Economics and the Problem of Allocating Airport Slots*. Boulder, CO: Westview Press.
Grossman, Sanford, and Joseph Stiglitz. 1980. "On the Impossibility of Informationally Efficient Markets." *American Economic Review* 70: 393–408.
Grossman, Sanford. 1989. *The Informational Role of Prices*. Cambridge, MA: MIT Press.
Gul, Faruk, and Wolfgang Pesendorfer. 2008. "The Case for a Mindless Economics." In A. Caplin and A. Schotter, eds., *The Foundations of positive and Normative Economics*, pp. 2–39. Oxford: Oxford University Press.
Hands, Wade. 1990. "Grunberg and Modigliani, Public Predictions, and New Classical Macroeconomics." *Research in the History of Economic Thought and Methodology* 7: 207–223.
Hands, Wade. 2015. "Orthodox and Heterodox Economics in Recent Economic Methodology." *Erasmus Journal for Philosophy and Economics* 8: 61–81.
Harcourt, Bernard. 2011. *The Illusion of Free Markets*. Cambridge, MA: Harvard University Press.
Harsanyi, John. 1967. "Games with Incomplete Information Played by Bayesian Players, Parts I–III." *Management Science* 14: 159–182, 320–334, 486–502.
Hartwell, R. M. 1995. *A History of the Mont Pelerin Society*. Indianapolis, IN: Liberty Fund.
Hayek, Friedrich. 1948. *Individualism and Economic Order*. Chicago: Regnery.
Hayek, Friedrich. 1952/1976. *The Sensory Order*. Chicago: University of Chicago Press.
Hayek, Friedrich. 1960. *The Constitution of Liberty*. Chicago: University of Chicago Press.

Hayek, Friedrich. 1967. *Studies in Philosophy, Politics and Economics.* New York: Simon & Schuster.

Hayek, Friedrich. 1978. *New Studies in Philosophy, Politics, Economics and the History of Ideas.* London: Routledge.

Hayek, Friedrich. 1982. "The Sensory Order after 25 Years." In W. Weimer and D. Palermo, eds., *Cognition and the Symbolic Process,* pp. 287–293. Hillsdale, NJ: Lawrence Erlbaum.

Hayek, Friedrich. 1988. *The Fatal Conceit.* Chicago: University of Chicago Press.

Heims, Steve. 1991. *The Cybernetics Group.* Cambridge, MA: MIT Press.

Helm, Leslie. 1994. "Nobel Puts a Spotlight on Game Theory." *Los Angeles Times,* October 19, 1994, D1.

Herfeld, Catherine. 2013. "Axiomatic Choice Theory Travelling Between Mathematical Formalism, Normative Choice Rules, and Psychological Measurement, 1944–1956." CHOPE Working Paper No. 2013-11.

Heukelom, Floris. 2014. *Behavioral Economics: A History.* New York: Cambridge University Press.

Heyck, Hunter. 2012. "Producing Reason." In Mark Solovey, ed., *Cold War Social Science,* pp. 99–116. New York: Palgrave Macmillan.

Heyck, Hunter. 2015. *The Age of System.* Baltimore: Johns Hopkins University Press.

Hidalgo, Cesar. 2015. *Why Information Grows.* New York: Basic Books.

Hildreth, Clifford. 1986. *The Cowles Commission in Chicago.* Berlin: Springer.

Hirshleifer, Jack. 1973. "Where are We in the Theory of Information?" *American Economic Review* 63(2): 31–39.

Hollinger, David. 1997. "The Disciplines and the Identity Debates, 1970–1995." *Daedalus* 126: 333–351.

Holmstrom, Bengt, Paul Milgrom, and Alvin Roth. 2002. Introduction. In Bengt Holmstrom, Paul Milgrom, and Alvin Roth, eds., *Game Theory in the Tradition of Bob Wilson,* Paper 1. Berkeley, CA: Bepress. At http://services.bepress.com/wilson/art1.

Hong, Harrison, Jeremy Stein, and Jialin Yu. 2007. "Simple Forecasts and Paradigm Shifts." *The Journal of Finance* 62(3): 1207–1242.

Horning, Rob. 2012. "Agents without Agency." *New Inquiry* No. 5. At http://thenewinquiry.com/essays/agents-without-agency/.

Hurwicz, Leonid. 1955. "Decentralized Resource Allocation." Cowles Commission Discussion Paper No. 2112.

Hurwicz, Leonid. 1960. "Optimality and Informational Efficiency in Resource Allocation Processes." In K. Arrow et al., eds., *Mathematical Methods in the Social Sciences, 1959,* pp. 27–46. Stanford, CA: Stanford University Press.

Hurwicz, Leonid. 1969. "On the Concept and Possibility of Information Decentralization." *American Economic Review* 59(2): 513–524.

Hurwicz, Leonid. 1971. "Centralization and Decentralization in Economic Processes." In A. Eckstein, ed., *Comparison of Economic Systems*, pp. 79–102. Berkeley: University of California Press.

Hurwicz, Leonid. 1972/1986. "On Informationally Decentralized Systems." In C. B. McGuire and R. Radner, eds., *Decision and Organization*, 2nd ed., pp. 297–336. Minneapolis: University of Minnesota Press.

Hurwicz, Leonid. 1977. "On the Dimensionality Requirements of Informationally Decentralized Pareto-Satisfactory Processes." In Kenneth Arrow and Leonid Hurwicz, eds., *Studies in Resource Allocation Processes*, pp. 413–424. New York: Cambridge University Press.

Hurwicz, Leonid. 1979. "On the Interaction between Information and Incentives in Organizations." In K. Krippendorff, ed., *Communication and Control in Society*, pp. 123–147. New York: Gordon and Breach.

Hurwicz, Leonid. 1986. "On Informational Decentralization and Efficiency of Resource Allocation Mechanisms." In S. Reiter, ed., *Studies in Mathematical Economics*, vol. 25, pp. 238–350. Washington, DC: Mathematical Association of America.

Hurwicz, Leonid. 1991. "'Economic Planning and the Knowledge Problem': A Comment." In John C. Wood and Ronald N. Woods, eds., *Friedrick A. Hayek: Critical Assessments*, vol. 4., pp. 83–88. New York: Routledge.

Hurwicz, Leonid. 2007. "But Who Will Guard the Guardians?" Bank of Sweden Prize lecture. At www.nobelprize.org/nobel_prizes/economic-sciences/laureates/2007/hurwicz_lecture.pdf.

Hurwicz, Leonid, and Stanley Reiter. 2006. *Designing Economic Mechanisms*. New York: Cambridge University Press.

Hurwicz, Leonid, and William Thomson. 1991. "Iterative Planning Procedures with a Finite Memory." In S. B. Dahia, ed., *Theoretical Foundations of Development Planning*, vol. 2, pp. 109–125. New Delhi: Concept Publishing.

Ingrao, Bruna, and Giorgio Israel. 1990. *The Invisible Hand*. Cambridge, MA: MIT Press.

Innocenti, Alessandro. 1995. "Oskar Morgenstern and the Heterodox Possibilities of Game Theory." *European Journal of the History of Economic Thought* 17: 205–227.

Jackson, Matthew. 1999. "The Non-existence of Equilibrium in Auctions with Two Dimensional Types." *Caltech Division of Humanities and Social Sciences Working Paper*: 228–277.

Jackson, Matthew. 2009. "Non-Existence of Equilibrium in Vickrey, Second-Price, and English Auctions." *Review of Economic Design* 13(1): 137–145.

Jaffe, Adam, and Josh Lerner. 2004. *Innovation and its Discontents*. Princeton, NJ: Princeton University Press.

Jehiel, Philippe, and Benny Moldovanu. 2001. "Efficient Design with Independent Valuations." *Econometrica* 69(5): 1237–1259.

Jones, Daniel Stedman. 2012. *Masters of the Universe*. Princeton: Princeton University Press.

Kagel, John, and Dan Levin. 2001. "Behavior in Multi-Unit Demand Auctions: Experiments with Uniform Price and Dynamic Vickrey Auctions." *Econometrica* 69(2): 413–454.

Kagel, John, and Dan Levin. 2009. "Implementing Efficient Multi-Object Auction Institutions: An Experimental Study of the Performance of Boundedly Rational Agents." *Games and Economic Behavior* 66(1): 221–237.

Kahlil, Elias. 2002. "Information, Knowledge and the Close of Friedrich Hayek's System." *Eastern Economic Journal* 28: 319–341.

Kauppinen, Ilkka. 2013. "Different Meanings of Knowledge as Commodity in the Context of Higher Education." *Critical Sociology*. doi: 10.1177/0896920512471218.

Kay, Lily. 2000. *Who Wrote the Book of Life?* Stanford, CA: Stanford University Press.

Khurana, Rakesh. 2010. *From Higher Aims to Hired Hands*. Princeton, NJ: Princeton University Press.

Kiesling, Lynne. 2015. "The Knowledge Problem." In P. Boettke, ed., *Oxford Handbook of Austrian Economics*, pp. 45–64. Oxford: Oxford University Press.

Klaes, Matthias, and Esther-Mirjam Sent. 2005. "A Conceptual history of the Emergence of Bounded Rationality." *History of Political Economy* 37: 27–59.

Klein, Daniel, ed. 2005. "Symposium on Information and Knowledge in Economics." *Economics Journal Watch* (2): 47–55.

Klein, Lawrence. 1991. "Econometric Contribution of the Cowles Commission." *Banca Nazionale del Lavoro Quarterly Review*. 177: 107–117.

Klemperer, Paul. 2004. *Auctions: Theory and Practice*. Princeton, NJ: Princeton University Press.

Klemperer, Paul. 2010. "The Product-Mix Auction: A New Auction Design for Differentiated Goods." *Journal of the European Economic Association* 8(2/3): 526–536.

Kline, Ronald. 2004. "What Is Information Theory a Theory Of? Boundary Work Among Scientists in the United States and Britain During the Cold War." In W. Boyd Rayward and Mary Ellen Bowden, eds., *The History and Heritage of Scientific and Technical Information Systems: Proceedings of the 2002 Conference, Chemical Heritage Foundation*, pp. 15–28. Medford, NJ: Information Today.

Kline, Ronald. 2015. *The Cybernetics Moment*. Baltimore: Johns Hopkins University Press.

Knight, Frank. 1940. "What Is Truth in Economics?" *Journal of Political Economy* 48: 1–32.

Knight, Frank. (1921) 1964. *Risk, Uncertainty and Profit*. New York: Augustus Kelley.

BIBLIOGRAPHY

Koopmans, Tjalling, ed. 1951. *Activity Analysis of Production and Distribution: Proceeding of a Conference*. Cowles Commission Monographs. New York: John Wiley.

Koopmans, Tjalling. 1957. *Three Essays on the State of Economic Science*. New York: McGraw-Hill. [reprint Augustus Kelley, 1991]

Kreps, David. 1990. *A Course in Microeconomic Theory*. Princeton, NJ: Princeton University Press.

Kreps, David. 1997. "Economics—the Current Position." In Thomas Bender and Carl Schorske, eds., *American Academic Culture in Transition*, pp. 77–104. Princeton, NJ: Princeton University Press.

Kripke, Saul. 1963. "A Semantical Analysis of Modal Logic I." *Zeitschrift fur Mathematische Logik und Grundlagen der Mathematik* (24): 323.

Krippner, Greta. 2011. *Capitalizing on Crisis: the political origins of the rise of finance*. Cambridge, MA: Harvard University Press.

Krishna, Vijay. 2002. *Auction Theory*. San Diego: Academic Press.

Kwerel, Evan. 2004. "Forward." In P. Milgrom, *Putting Auction Theory to Work*, pp. xv–xxii. New York: Cambridge University Press.

Kwerel, Evan, and Gregory Rosston. 2000. "An Insiders' View of the FCC Spectrum Auctions." *Journal of Regulatory Economics* 17(3): 253–289.

Labaton, Stephen, and Simon Romero. 2001. "Wireless Giants Won F.C.C. Auction Unfairly, Critics Say." *New York Times*, February 12, 2001.

Laffont, Jean-Jacques, and Jean Tirole. 1993. *A Theory of Incentives in Procurement and Regulation*. Cambridge, MA: MIT Press.

Lagueux, Maurice. 2010. *Rationality and Explanation in Economics*. London: Routledge

Lamberton, Donald. 1998. "Information Economics Research: Points of Departure." *Information Economics and Policy* 10: 325–330.

Lange, Oskar and Taylor, Fred. 1964. *On the Economic Theory of Socialism*. New York: McGraw Hill.

Lavoie, Don. 1985. *Rivalry and Central Planning*. New York: Cambridge University Press.

Lavoie, Don. 1986. "The Market as a Procedure for Discovery and Conveyance of Inarticulate Knowledge." *Comparative Economic Studies* 28(1): 1–19.

Ledyard, John. 1993. "The Design of Coordination Mechanisms and Organizational Computing." *Journal of Organizational Computing* 3(1): 121–134.

Ledyard, John, David Porter, and Antonio Rangel. 1997. "Experiments Testing Multiobject Allocation Mechanisms." *Journal of Economics and Management Strategy* 6(3): 639–675.

Lee, Kyu Sang. 2006. "Mechanism Design Theory Embodying an Algorithm-Centered Vision of Markets/Organizations/Institutions." *History of Political Economy* 38 (Suppl.): 283–304.

Lee, Kyu Sang. 2015. "Mechanism Designers in Alliance: A Portrayal of a Scholarly Network in Support of Experimental Economics." Working Paper. [reprint Kyu Sang Lee, 2016]

Leppälä, Samuli. 2015. "Economic Analysis of Knowledge: The History of Thought." *Journal of Economic Surveys* 29: 263–286.

Lerner, Abba. 1944. *The Economics of Control*. New York: Macmillan.

Levine, David, and Steven Lippman, eds. 1995. *The Economics of Information*. Cheltenham: Edward Elgar.

Levy, Arnon. 2011. "Information in Biology: A Fictionalist Account." *Nous* 45: 640–657.

Lloyd, Seth. 2006. *Programming the Universe*. New York: Alfred Knopf.

Machlup, Fritz. 1962. *The Production and Distribution of Knowledge in the U.S.* Princeton, NJ: Princeton University Press.

Machlup, Fritz. 1980. *Knowledge: Its Creation, Distribution and Economic Significance*, 3 vols. Princeton, NJ: Princeton University Press.

Machlup, Fritz, and Una Mansfield, eds. 1983. *The Study of Information*. New York: John Wiley.

Machol, Robert, ed. 1960. *Information and Decision Processes*. New York: McGraw-Hill.

MacKenzie, Donald, Fabien Muniesa, and Lucia Siu, eds. 2007. *Do Economists Make Markets? On the Performativity of Economics*. Princeton, NJ: Princeton University Press.

Mandelbaum, Eric. 2015. "Associationist Theories of Thought." At http://plato.stanford.edu/entries/associationist-thought/.

Mandelbrot, Benoit. 1953. *Contributions à la théorie mathématique des jeux de communications*. Paris: Institut de Statistiques de l'Université de Paris.

Marschak, Jacob. 1959. "Remarks on the Economics of Information." In Marschak, Jacob, ed., *Economic, Information, Decision and Prediction*, pp. 91–117. Dordrecht: Reidel.

Marschak, Jacob. 1971. "Optimal Symbol Processing: A Problem in Individual and Social Economics." *Behavioral Science* 16: 202–217.

Marschak, Jacob. 1974. *Economic Information, Decision and Prediction*, 2 vols. Dordrecht: Reidel.

Martin, Randy. 2015. "Coming Up Short: Knowledge Limits and the Decomposition of the Professional Managerial Class." *International Critical Thought* 5(1): 95–110.

Maskin, Eric. 1992. "Auctions and Privatization." In H. Siebert, ed. *Privatization*, pp. 115–136. Kiel: Institut fur Weltwirtschaften der Universität Kiel.

Maskin, Eric. 2015. "Friedrich von Hayek and Mechanism Design." *Review of Austrian Economics* 28(3): 247–252.

McAfee, R. P. 1993a. "Auction Design for Personal Communications Services." Comments of PacTel, PP Docket No. 93-253.

McAfee, R. P. 1993b. "Auction Design for Personal Communications Services: Reply Comments." PacTel Reply Comments, PP Docket No. 93-253.

McAfee, R. P., and John McMillan. 1987. "Auctions and Bidding." *Journal of Economic Literature* 25(2): 699–738.

McAfee, R. P., and John McMillan. 1996. "Analyzing the Airwaves Auction." *Journal of Economic Perspectives* 10(1): 159–175.

McCabe, Kevin, Stephen Rassenti, and Vernon Smith. 1991. "Smart Computer-Assisted Markets." *Science* 254(5031): 534–538.

McCabe, Kevin, Stephen Rassenti, and Vernon Smith. 1999. "Designing Auction Institutions for Exchange." *IIE Transactions* 31(9): 803–811.

McGlynn, Aidan. 2014. *Knowledge First?* London: Palgrave.

McGrayne, Sharon. 2011. *The Theory that Would Not Die*. New Haven, CT: Yale University Press.

McGuire, C. B., and Roy Radner, eds. 1986. *Decision and Organization*. Minneapolis: University of Minnesota Press.

McKee, Michael. 1990. "Review of *The Allocation of Scarce Resources* by Grether, Isaac, and Plott." *Journal of Economic Literature* 28(2): 711–713.

McMillan, John. 1994. "Selling Spectrum Rights." *Journal of Economic Perspectives* 8(3): 145–162.

McMillan, John. 1995. "Why Auction the Spectrum?" *Telecommunications Policy* 19(3): 191–199.

McMillan, John. 2002. *Reinventing the Bazaar*. New York: W.W. Norton.

McMillan, John. 2003. "Market Design: The Policy Uses of Theory." *AEA Papers and Proceedings* 93(2): 139–144.

McMillan, John. 2004. "Using Markets to Help Solve Public Problems." In Takatoshi Ito and Anne O. Krueger, eds., *Governance, Regulation, and Privatization in the Asia-Pacific Region*, pp. 73–92. Chicago: University of Chicago Press.

McMillan, John, Michael Rothschild, and Robert Wilson. 1997. "Introduction." *Journal of Economics and Management Strategy* 6(3): 425–430.

Mehrling, Perry. 2005. *Fischer Black and the Revolutionary Idea of Finance*. New York: John Wiley.

Meister, Alan. 1999. "Evaluating the Performance of the Spectrum Auctions: A Case Study of the PCS Auctions." PhD dissertation, Department of Economics, University of California, Irvine.

Mikoucheva, A., and K. Sonin. 2004. "Information Revelation and Efficiency in Auctions." *Economics Letters* 83(3): 277–284.

Milgrom, Paul. 2004. *Putting Auction Theory to Work*. New York: Cambridge University Press.

Milgrom, Paul, and John Roberts. 1986. "Relying on the Information of Interested Parties." *RAND Journal of Economics* 17: 18–32.

Milgrom, Paul, and Nancy Stokey. 1982. "Information, Trade and Common Knowledge." *Journal of Economic Theory* 26: 17–27.

Milgrom, Paul, and Robert Weber. 1982. "A Theory of Auctions and Competitive Bidding." *Econometrica* 50(5): 1089–1122.

Milgrom, Paul, and Robert Wilson. 1993a. "Affidavit of Paul R. Milgrom and Robert B. Wilson." Comments of PacBell, PP Docket No. 93-253.

Milgrom, Paul, and Robert Wilson. 1993b. "Replies to Comments on PCS Auction Design." Comments of PacBell, PP Docket No. 93-253.

Mirowski, Philip. 1989. *More Heat than Light*. New York: Cambridge University Press.

Mirowski, Philip, ed. 1994. *Edgeworth on Chance, Economic Hazard, and Statistics*. Totawa: Rowman & Littlefield.

Mirowski, Philip. 1998. "Economics, Science and Knowledge." *Tradition and Discovery* 25: 29–42.

Mirowski, Philip. 2002. *Machine Dreams*. New York: Cambridge University Press.

Mirowski, Philip. 2006. "Twelve Theses Concerning the History of Postwar Neoclassical Price Theory." In Wade Hands and Philip Mirowski, eds., *Agreement on Demand*, pp. 343–379. Durham, NC: Duke University Press.

Mirowski, Philip. 2007. "Markets Come to Bits." *Journal of Economic Behavior and Organization* 63(2): 209–242.

Mirowski, Philip. 2009. "Why There Is (as yet) No Such Thing as an Economics of Knowledge." In Harold Kincaid and Don Ross, eds., *Oxford Handbook of the Philosophy of Economics*, pp. 99–156. Oxford: Oxford University Press.

Mirowski, Philip. 2011. *ScienceMart: privatizing American science*. Cambridge, MA: Harvard University Press.

Mirowski, Philip. 2012. "The Cowles Anti-Keynesians." In Pedro Duarte and Gilberto Lima, eds., *Microfoundations Reconsidered: The Relationship of Micro and Macroeconomics in Historical Perspective*, pp. 131–177. London: Edward Elgar.

Mirowski, Philip. 2013. *Never Let a Serious Crisis Go to Waste*. New York: Verso.

Mirowski, Philip. 2015. The Neoliberal Ersatz Nobel in Economics." Paper presented to Road to Mont Pèlerin 2 conference.

Mirowski, Philip, and Edward Nik-Khah. 2007. "Markets Made Flesh: Performativity, and a Problem in Science Studies, augmented with Consideration of the FCC Auctions." In Donald MacKenzie et al., *Do Economists Make Markets? On the Performativity of Economics*, pp. 190–225. Princeton, NJ: Princeton University Press.

Mirowski, Philip, and Edward Nik-Khah. 2008. "Command Performance: Exploring What STS Thinks it Takes to Build a Market." In Trevor Pinch and Richard Swedberg, eds., *Living in a Material World: Economic Sociology Meets Science and Technology Studies*, pp. 89–128. Cambridge MA: MIT Press.

BIBLIOGRAPHY

Mirowski, Philip, and Dieter Plehwe, eds. 2009. *The Road from Mont Pèlerin: The Making of the Neoliberal Thought Collective*. Cambridge, MA: Harvard University Press.

Mirowski, Philip, and Roy Weintraub. 1994. "The Pure and the Applied: Bourbakism Comes to Mathematical Economics." *Science in Context* 7: 245–272.

Morgenstern, Oskar. 1935. "Perfect Foresight and Economic Equilibrium." *Zeitschrift fur Nationalokonomie* 6 (originally in German). [reprint Andrew Schotter, ed. 1976]

Mount, Kenneth, and Stanley Reiter. 1974. "The Informational Size of Message Spaces." *Journal of Economic Theory* 8(2): 161–192.

Mount, Kenneth, and Stanley Reiter. 2002. *Computation and Complexity in Economic Behavior and Organization*. New York: Cambridge University Press.

Murray, James. 2002. *Wireless Nation*. Cambridge, MA: Perseus.

Myerson, Roger. 2004. "Comments on 'Games with Incomplete information Played by Bayesian Players.'" *Management Science* 50: 1818–1824.

Myerson, Roger. 2008. "Mechanism Design." In Steven N. Durlauf and Lawrence E. Blume, eds., *The New Palgrave Dictionary of Economics Online*, 2nd ed. Palgrave Macmillan. At http://www.dictionaryofeconomics.com/article?id=pde2008_M000132. doi:10.1057/9780230226203.1078.

Myerson, Roger. 2009. "Fundamental Theory of Institutions: A Lecture in Honor of Leo Hurwicz." *Review of Economic Design* 13: 59–75.

Nalebuff, Barry, and Jeremy Bulow. 1993a. "Designing the PCS Auction." Comments of Bell Atlantic, FCC PP Docket No. 93-253.

Nalebuff, Barry, and Jeremy Bulow. 1993b. "Response to PCS Auction Design Proposals." Reply Comments of Bell Atlantic, FCC PP Docket No. 93-253.

Niehans, Jürg. 1990. *A History of Economic Theory*. Baltimore: Johns Hopkins University Press.

Nik-Khah, Edward. 2005. "Designs on the Mechanism." PhD dissertation, University of Notre Dame, South Bend, IN.

Nik-Khah, Edward. 2008. "A Tale of Two Auctions." *Journal of Institutional Economics* 4(3): 73–97.

Nik-Khah, Edward. 2011. "George Stigler, the Graduate School of Business, and the Pillars of the Chicago School." In Robert Van Horn, Philip Mirowski, and Tom Stapleford, eds., *Building Chicago Economics: New Perspectives on the History of America's Most Powerful Economics Program*, pp. 116–147. New York: Cambridge University Press.

Nik-Khah, Edward, and Robert Van Horn. 2015. "The Ascendancy of Chicago Neoliberalism." In S. Springer, K. Birch, and J. MacLeavy, eds., *Handbook of Neoliberalism*, pp. 27–38. London: Routledge.

Noam, Eli. 1995. "Taking the Next Step Beyond Spectrum Auctions." Unpublished manuscript. At http://www.columbia.edu/dlc/wp/citi/citinoam21.html.

North, Douglas. 1977. "Markets and other Allocation Systems in History." *Journal of European Economic History* 6(3): 703–716.
Oguz, Fuat. 2010. "Hayek on Tacit Knowledge." *Journal of Institutional Economics* 6: 145–166.
Pardo-Guerra, Juan. 2010. "Computerising Gentlemen: The Automation of the London Stock Exchange, 1945–1995." PhD thesis, STS, University of Edinburgh.
Pardo-Guerra, Juan, and Donald MacKenzie. 2014. "The Politics of Fragmentation: Liberalism, Market Equality, and the Technological Reconfiguration of American Finance." Working paper. At https://pardoguerra. files.wordpress.com/2014/03/pardo-guerra-and-mackenzie-the-politics-of-fragmentation1.pdf.
Pareto, Vilfredo. 1981. Manuel d'économie politique. Geneva: Droz.
Paulson, Henry. 2010. *On the Brink*. New York: Business Plus.
Plant, Raymond. 2010. *The Neoliberal State*. Oxford: Oxford University Press.
Plott, Charles. 1979. "The Application of Laboratory Experimental Methods to Public Choice." In C. Russell, ed., *Collective Decision Making: Applications from Public Choice Theory*, pp. 137–160. Baltimore: Johns Hopkins University Press.
Plott, Charles. 1994. "Market Architectures, Institutional Landscapes, and Testbed Experiments." *Economic Theory* 4(1): 3–10.
Plott, Charles. 1997. "Laboratory Experimental Testbeds: Application to the PCS Auction." *Journal of Economics and Management Strategy* 6(3): 605–638.
Plott, Charles. 2001. *Market Institutions and Price Discovery*. Northampton, MA: Edward Elgar.
Plott, Charles. 2005. "Dimensions of Parallelism: Some Policy Applications of Experimental Methods." In A. Roth, ed., *Laboratory Experimentation in Economics: Six Points of View*, pp. 193–219. New York: Cambridge University Press.
Plott, Charles. 2014. "Public Choice and the Development of Modern Laboratory Experimental Methods in Economics and Political Science." *Constitutional Political Economy* 25: 331–353.
Porter, David, Stephen Rassenti, Anil Roopnarine, and Vernon Smith. 2003. "Combinatorial Auction Design." *Proceedings of the National Academy of Sciences* 100(19): 11153–11157.
Prendergast, Renee. 2007. "Knowledge and Information in Economics: What Did the Classicals Know?" *History of Political Economy* (39): 679–712.
Purcell, Edward. 1973. *The Crisis of Democratic Theory: Scientific Naturalism and the Problem of Value*. Lexington: University of Kentucky Press.
Quastler, Henry. 1955. *Information Theory in Psychology: Problems and Methods*. Glencoe, IL: Free Press.

Quinn, Sarah. 2010. "Government Policy, Housing, and the Origins of Securitization." PhD dissertation, University of California, Berkeley.

Quirk, James, and Rubin Saposnik. 1968. *Introduction to General Equilibrium Theory and Welfare Economics*. New York: McGraw-Hill.

Radner, Roy. 1968. "Competitive Equilibrium Under Uncertainty." *Econometrica*. 36: 31–58.

Raiffa, Howard. 2002. "Tribute to Robert Wilson on his 65th Birthday." In Bengt Holmstrom, Paul Milgrom, and Alvin Roth, eds., *Game Theory in the Tradition of Bob Wilson*, Paper 2. Berkeley, CA: Bepress. At http://services.bepress.com/wilson/art2.

Rassenti, Stephen. 1982. "Zero/One Decision Problems with Multiple Resources and Constraints: Algorithms and Applications." PhD dissertation, University of Arizona.

Rassenti, Stephen, Vernon Smith, and Robert Bulfin. 1982. "A Combinatorial Auction Mechanism for Airport Time Slot Allocation." *Bell Journal of Economics* 13(2): 402–417.

Reiter, Stanley. 1977. "Information and Performance in the (New)^2 Welfare Economics." *American Economic Review* 67: 226–234.

Reiter, Stanley. 2001. "Coordination of Economic Activity—An Example." *Review of Economic Design* 6: 263–288.

Reiter, Stanley. 2009. "Two Topics in Leo Hurwicz's Research." *Review of Economic Design* 13: 3–6.

Reiter, Stanley and Maroulis. 2008. "Stable Processes of Exchange." *Journal of Mathematical Economics*. 44: 1398–1412.

Reiter, Stanley, and Kenneth Mount. 1974. "The Informational Size of Message Spaces." *Journal of Economic Theory* 8: 161–192.

Reiter, Stanley, and Kenneth Mount. 2002. *Computational Complexity in Economic Behavior and Organization*. New York: Cambridge University Press.

Reiter, Stanley, Kenneth Arrow, Lance Davis, Paul Dimaggio, Mark Granovetter, Jerry Green, Theodore Groves, Michael Hannan, Andrew Postlewaite, Roy Radner, Karl Shell, and Leonid Hurwicz. 1989. "Markets and Organizations." In R. D. Luce, Neil Smelser, and Dean Gernstein, eds., *Leading Edges in the Social and Behavioral Sciences*, pp. 283–326. New York: Russell Sage.

Rizvi, S. Abu Turab. 1994. "The Microfoundations Project in General Equilibrium Theory." *Cambridge Journal of Economics* 18: 357–377.

Rizvi, S. Abu Turab. 1998. "Responses to Arbitrariness in Contemporary Economics." In John Davis, ed., *New Economics and its History*, Supplement to vol. 29, *History of Political Economy*, pp. 273–288. Durham, NC: Duke University Press.

Rodrik, Dani. 2014. "When Ideas Trump Interests." *Journal of Economic Perspectives* (28): 189–208.

Romer, Paul. 1990. "Endogenous Technical Change." *Journal of Political Economy.* 98: S71-S102.

Romer, Paul. 2015. "Mathiness in the Theory of Economic Growth." *American Economic Review: Papers and proceedings* 105(5): 89–93.

Ropke, Wilhelm. 1949. *Civitas humana.* London: Hodge.

Rosenbaum, Eckehard. 2000. "What Is a Market?" *Review of Social Economy* 58(4): 455–482.

Rothschild, Michael, Joseph Stiglitz. 1978. "Equilibrium in Competitive Insurance Markets: An Essay on the Economics of Imperfect Information." In Peter Diamond and Michael Rothschild, eds., *Uncertainty in Economics: Readings and Exercises,* pp. 259–280. New York: Academic Press.

Roth, Alvin. 2002a. "The Economist as Engineer: Game Theory, Experimentation and Computation as Tools for Design Economics." *Econometrica* 70(4): 1341–1378.

Roth, Alvin. 2002b. "Preface to 'The Redesign of the Matching Market for American Physicians.'" In Bengt Holmstrom, Paul Milgrom, and Alvin Roth, eds., *Game Theory in the Tradition of Bob Wilson,* Paper 15. Berkeley, CA: Bepress. At www.bepress.com/wilson/art15.

Roth, Alvin. 2015. *Who Gets What—and Why.* Boston: Houghton Mifflin Harcourt.

Rothkopf, Michael. 1969. "A Model of Rational Competitive Bidding." *Management Science* 15(7): 362–373.

Rothkopf, Michael. 2000. "Tales from a Nonstandard Career in Operations Research." RUTCOR Research Report 48-2000.

Royal Swedish Academy of Sciences. 2007. "Scientific Background on Mechanism Design Theory." At www.nobelprize.org/nobel_prizes/economic-sciences/laureates/2007/advanced-economicsciences2007.pdf

Rumsfeld, Donald. 2010. "Press Conference at NATO HQ NATO Speeches." At www.nato.int/docu/speech/2002/s020606g.htm.

Samuelson, Larry. 2004. "Modelling Knowledge in Economic Analysis." *Journal of Economic Literature* (42): 367–403.

Samuelson, Paul. 1954. "The Pure Theory of Public Expenditure." *Review of Economics and Statistics.* 36: 387-389.

Samuelson, Paul. 2009. "A Few Remembrances of Friedrich von Hayek." *Journal of Economic Behavior and Organization* 69: 1–4.

Sapienza, Paola, and Luigi Zingales. 2013. "Economic Experts vs. Average Americans." *American Economic Review, Papers and Proceedings* 103(3): 636–642.

Scheall, Scott. 2015. "A Brief Note Concerning Hayek's Non-standard Conception of Knowledge." *Review of Austrian Economics* 10(1007).

Schelling, Thomas. 1962. "Forward" in Roberta Wohlstetter, *Pearl Harbor: Warning and Decision.* Stanford: Stanford University Press.

Schiller, Dan. 1988. "How to Think about Information." In V. Mosco and J. Wasko, eds., *The Political Economy of Information*, pp. 27–43. Madison: University of Wisconsin Press.

Sent, Esther-Mirjam. 1998. *The Evolving Rationality of Rational Expectations Theory*. New York: Cambridge University Press.

Sent, Esther-Mirjam. 2001. "Sent Simulating Simon Simulating Scientists." *Studies in the History and Philosophy of Science* 32: 479–500.

Sent, Esther-Mirjam. 2004. "The Legacy of Herbert Simon in Game Theory." *Journal of Economic Behavior and Organization* 53: 303–317.

Shannon, Claude. 1948. "The Mathematical Theory of Communication." *Bell System Technical Journal* 27: 379–423, 623–656.

Shannon, Claude. 1956. "The Bandwagon." *IRE Trans IT* 2: 3.

Shapiro, Carl, and Hal Varian. 1999. *Information Rules*. Cambridge, MA: Harvard Business School Press.

Shi, Yenfei. 2001. *The Economics of Scientific Knowledge*. Cheltenham: Edward Elgar.

Simon, Herbert. 1968/1981. *The Sciences of the Artificial*. Cambridge, MA: MIT Press.

Simon, Herbert. 1978. "On How to Decide What to Do." *Bell Journal of Economics* 9: 494–507.

Simon, Herbert. 1991. *Models of my Life*. New York: Basic Books.

Skarbek, David. 2009. "F.A. Hayek's Influence on Nobel Prize Winners." *Review of Austrian Economics* 22: 109–112.

Smith, Vernon. 1991. *Papers in Experimental Economics*. New York: Cambridge University Press.

Smith, Vernon. 2001. "From Old Issues to New Directions in Experimental Psychology and Economics." *Behavioral and Brain Sciences*. 24: 428–429.

Smith, Vernon. 2006. Foreword. In Peter Cramton, Yoav Shoham, and Richard Steinberg, eds., *Combinatorial Auctions*, pp. xi–xv. Cambridge, MA: MIT Press.

Smith, Vernon. 2008a. *Discovery—A Memoir*. Bloomington, IN: AuthorHouse.

Smith, Vernon. 2008b. *Rationality in Economics: Constructivist and Ecological Forms*. New York: Cambridge University Press.

Smith, Vernon. 2010. "Theory and Experiment: What Are the Questions?" *Journal of Economic Behavior and Organization* 73(1): 3–15.

Smith, Vernon. 2015. "Discovery Processes, Science, and 'Knowledge-how:' Competition as a Discovery Procedure in the Laboratory." *Review of Austrian Economics* 28(3): 237–245.

Solovey, Mark. 2013. *Shaky Foundations: The Politics-Patronage-Social Science Nexus in Cold War America*. New Brunswick, NJ: Rutgers University Press.

Solow, Robert. 2012. "Hayek, Friedman and the Illusions of Conservative Economics." *New Republic*, November 16.

BIBLIOGRAPHY

Sorenson, Roy. 1988. *Blindspots*. Oxford: Clarendon Press.

Sorkin, Andrew. 2009. *Too Big to Fail*. New York: Viking.

Spence, Michael, Richard Zeckhauser. 1978. "Insurance, Information, and Individual Action." In Peter Diamond and Michael Rothschild, eds., *Uncertainty in Economics: Readings and Exercises*, pp. 335–343. New York: Academic Press.

Stigler, George. 1961. "The Economics of Information." *Journal of Political Economy* 69: 213–225.

Stiglitz, Joseph. 1985. "Information and Economic Analysis: A Perspective." *Economic Journal* 75 (Suppl.): 21–41.

Stiglitz, Joseph. 2000. "The Contributions of the Theory of Information to 20th Century Economics." *Quarterly Journal of Economics* 140: 1441–1478.

Stiglitz, Joseph. 2003. "Information and the Change in Paradigm in Economics." In R. Arnott, B. Greenwald, R. Kanbur, and B. Nalebuff, eds., *Economics in an Imperfect World*, pp. 569–639. Cambridge, MA: MIT Press.

Stiglitz, Joseph. 2009. *Selected Works*. Oxford: Oxford University Press.

Stiglitz, Joseph. 2010a. *Freefall*. New York: Norton.

Stiglitz, Joseph. 2010b. "The Non-existent Hand." *London Review of Books*, April 22.

Stiglitz, Joseph. 2010c. "An Agenda for Reforming Economic Theory." Manuscript distributed at INET Conference, King's College, Cambridge, April.

Stiglitz, Joseph. 2011. "Rethinking Macroeconomics: What Failed, and How to Repair It." *Journal of the European Economic Association* 9: 591–645.

Sunder, Shyam. 2004. "Markets as Artifacts." In M. Augier and J. March, eds., *Models of a Man: Essays in Memory of Herbert Simon*, pp. 501–520. Cambridge, MA: MIT Press.

Svorenčík, Andrej, and Harro Maas. 2016. *The Making of Experimental Economics: Witness Seminar on the Emergence of a Field*. New York: Springer.

Swagel, Phillip. 2009. "The Financial Crisis: An Inside View." *Brookings Papers on Economic Activity*, Spring 2009: 1–63.

Taylor, R. G. 1998. *Models of Computation and Formal Languages*. New York: Oxford University Press.

Thelen, Jennifer. 1995. "Milgrom's Progress." *The Recorder*, May 8.

Thomas, Will. 2015. *Rational Action*. Cambridge, MA: MIT Press.

Tkacik, Maureen. 2010. "Journals of the Crisis Year." *Baffler* 18.

Van Damme, Eric. 1998. "On the State of the Art in Game Theory: An Interview with Robert Aumann." *Games and Economic Behavior* 24(1–2): 181–210.

Van Horn, R., and P. Mirowski. 2009. "The Rise of the Chicago School of Economics and the Birth of Neoliberalism." In P. Mirowski and D. Plehwe, eds., *The Road from Mont Pèlerin*, pp. 139–178. Cambridge, MA: Harvard University Press.

Van Horn, Robert, Mirowski, Philip and Thomas Stapleford, eds. 2011. *Building Chicago Economics*. New York: Cambridge University Press.

Varian, Hal. 2002. "A New Economy with No New Economics." *New York Times*, January 17. At www.nytimes.com/2002/01/17/business/17SCEN.html.

Veblen, Thorstein. 1930. *Place of Science in Modern Civilization*. New York: Viking.

Vickrey, William. 1960. "Utility, Strategy, and Social Decision Rules." *Quarterly Journal of Economics* 74(4): 507–535.

Vickrey, William. 1961. "Counterspeculation, Auctions and Competitive Sealed Tenders." *Journal of Finance* 16: 8–37.

von Foerster, H., Mead, M., and Teuber, H. L., eds. 1952. *Cybernetics: Circular Causal and Feedback Mechanisms in Biological and Social Systems. Transactions of the Eighth Conference*. New York: Josiah Macy, Jr. Foundation.

Vulkan, Nir, Alvin Roth, and Zvika Neeman. 2012. *The Handbook of Market Design*. New York: Oxford University Press.

Walker, Donald. 2001. "A Factual Account of the Functioning of the Nineteenth-Century Paris Bourse." *European Journal for the History of Economic Thought* 8(2): 186–207.

Warsh, David. 1993. *Economic Principles*. New York: Free Press.

Warsh, David. 2006. *Knowledge and the Wealth of Nations*. New York: W.W. Norton.

Weber, Robert. 1993a. "Comments on FCC 93-455." Comments of TDS, PP Docket No. 93-253.

Weber, Robert. 1993b. "Reply to Comments on FCC 93-455." Reply Comments of TDS, PP Docket No. 93-253.

Weintraub, E. Roy. 1991. *Stabilizing Dynamics*. New York: Cambridge University Press.

Wilkinson, Nick, and Matthias Klaes. 2012. *An Introduction to Behavioral Economics*. New York: Palgrave Macmillan.

Wilson, Robert. 1969a. "Arrow's Possibility Theorem for Vote Trading." In *Mathematical Theory of Committees and Elections*, pp. 26–39. Vienna: Institute for Advanced Studies.

Wilson, Robert. 1969b. "The Role of Uncertainty and the Value of Logrolling in Collective Choice Processes." In G. Guilbaud, ed., *La Decision: Agregation et Dynamique des Ordres de Preference*, pp. 309–315. Paris: Centre National de la Recherche Scientifique.

Wilson, Robert. 1969c. "The Structure of Incentives for Decentralization under Uncertainty." In G. Guilbaud, ed., *La Decision: Agregation et Dynamique des Ordres de Preference*, pp. 287–307. Paris: Centre National de la Recherche Scientifique.

Wilson, Robert. 1996. "John Harsanyi and the Economics of Information." *Games and Economic Behavior* 14(2): 296–298.

INDEX

Airline Deregulation Act, 139–140
Airtouch, 214
Akerlof, George, 38–39, 106, 163
Ambler, Eric, 9–10
American Institutionalism, 77
Ames, Ed, 95
Arizona, University of, 124, 184–185
　Economics Department, 139, 201, 205
Arrow, Kenneth, 92–94, 126, 131, 144, 170
artificial intelligence, 91, 109
Ashby, Ross, 84
Ashenfelter, Orley, 180
associationist psychology, 68
asymmetric information, 36, 40, 73, 93,
　99, 106, 109, 116, 197
auctionomics, 219–220
auctions, 140–142, 171–181, 186, 200,
　211–215, 217, 219–220, 221, 224,
　226–231, 236, 260n17, 260n21
　clock, 219–220, 227–229
　combinatorial, 121, 213
　common value, 190
　double, 120–123
　Dutch, 157, 171
　English, 171, 174, 179–180, 219, 264n6
　Milgrom-Wilson, 219
　open book, 140, 175, 226
　reverse, 224, 228
　sealed bid, 140, 171, 219–220, 226–227
　Vickrey, 141

Aumann, Robert, 99, 110, 113, 115–116
Austrian School of Economics, 25, 54, 60,
　73–75, 194–195
Ausubel, Lawrence, 223–224,
　227–229, 231

Barro, Robert, 7
Bayesian inductive inference, 100, 111,
　115–116
Bayesian rationality, 128
Bayes-Nash school, 100, 115–116, 149,
　158–159, 172–182, 190, 192,
　197–198, 200, 203–206,
　211–219
Bear Stearns, 222
Becker, Gary, 52, 55, 104–105
behavioral economics, 7, 18, 21–22, 36, 83,
　106, 128–129, 160
Bell Atlantic, 214
Bernanke, Ben, 222–224, 227, 231
bit (unit of information), 47, 107
Blackwell, David, 98–99, 108–109
Blackwell Program, 83–84, 99–100, 102,
　108–112, 136, 151, 152, 157,
　159, 204
block universe, 109–110, 112, 115
Boldyrev, Ivan, 149
Boulding, Kenneth, 23–24, 38
Buchanan, James, 139
Bulfin, Robert, 142, 183–184

INDEX

Bulow, Jeremy, 214
Bush, President George W., 231

Caltech, 124, 139–140, 201
Capen, Ed, 177
Carnegie Mellon, 91, 122
Carter, President Jimmy, 138
Center for Mathematical Studies in Economics and Management Science, 201
Chamberlin, Edward, 120
channel capacity, 47, 88–89, 135, 166
Chicago, University of, 61
  Committee on Social Thought, 68, 75
  School of Economics, 3, 54, 61, 75, 159
Chicago Booth Financial Trust Index, 3
choice under uncertainty, 111
Citigroup, 231
Clark Medal, 6, 118
Coase, Ronald, 52, 107, 218
Coase theorem, 107
Columbia University, 170
Committee on Basic Research in the Behavioral and Social Sciences, 125
commodity space, 103–104, 166
common knowledge, 110, 115–116, 174
common values, 180
common-value model, 198
computer, 27, 28, 45, 47, 49, 79, 90–92, 94, 96, 111, 118, 145–146, 152, 163, 185, 205
consensus theory of truth, 4, 5
Constant, Benjamin, 58
constructivism, 96
Cowles, Alfred, 73, 74
Cowles Commission, 60–61, 73–82, 87, 89–96, 98–100, 101–102, 118, 124, 129, 133–134, 161–162, 178, 249n1
Cramton, Peter, 172, 223–224, 227–229, 231
cryptography, 46
cybernetics, 47, 79–80, 134–135
Cybernomics, 205

Debreu, Gerard, 104
decentralization, 88, 134–136, 166, 172, 201–202, 250–251

decision theory, 16, 18, 28, 40, 47, 49, 76, 83–84, 90, 93, 172
DeMartino, George, 6
democracy, 8–9, 27, 29
deregulation, 138–142
designer markets, 184
Director, Aaron, 61

econometrics, 76, 80
economic experiments, 97, 118–121, 141
Economic Science Association, 201
efficiency, 123, 211, 228
efficient markets hypothesis, 147
electromagnetic spectrum licenses, 178, 205, 207–220
engineering economics, 122
entropy, 45–46, 84, 104, 248n3
epistemology, 2, 4, 8, 16, 23–24, 33–34, 70, 113, 152
Epstein, Gerry, 6
equilibrium models, 120, 168, 190
evolutionary psychology, 17, 224
experimentalist school, 119–121, 147, 150, 159, 181–182, 183–192, 198, 200, 203, 204, 205, 208, 212–213, 215, 220, 238

FAA (Federal Aviation Administration), 140, 205
Fannie Mae, 146
FCC (Federal Communications Commission), 207–221, 240–241
Ferejohn, John, 139
financial crisis of 2008, 4, 17, 103, 221–222, 232, 236
Fiorina, Morris, 139
Ford Foundation, 172
Foucault, Michel, 8–9, 33–34, 55
Freddie Mac, 146
freedom, 56
Friedman, Milton, 56, 61, 95, 235–236

Gale, David, 87, 189
game theory, 40, 91, 111, 113, 137, 156, 167, 181, 204, 208–215, 238
George Mason University, 195, 261n5
Gigerenzer, Gerd, 20

INDEX

Giocoli, Nicola, 22–27, 128
Gleick, James, 32
Gode, Dan, 122–123, 188
Gödel's Theorem, 19
Goldman Sachs, 231
Gorton, Gary, 33
Great Depression, the, 53
Green, Jerry, 132
Grether, David, 138–140
Groves, Theodore, 202

habit psychology, 26
Harcourt, Bernard, 58
Harsanyi, John, 40, 112–116, 172, 174, 197
Harsanyi-Nash Program, 112–116
Harvard Business School, 171–172
Hayek, Friedrich, 13–14, 52, 54–57, 61–65, 66–67, 73, 75, 132, 134, 136–137, 152–154, 157, 161, 168, 193–205
Hayek hypothesis, 120–121, 122, 198
Heukelom, Floris, 128
Heyck, Hunter, 27–28, 76, 124
Hirschman, Albert, 33
history and philosophy of science, 237
Hollinger, David, 33–34
"Homo economicus," 17, 22, 34, 55
Horning, Rob, 9–10, 12–13
human capital, 105
Hurwicz, Leonid, 85–89, 98, 124, 130, 131–137, 149–150, 151, 154, 167–168, 194–195, 202–203
Hutchins, Robert, 74

incentive compatibility, 80, 132, 135, 142, 167, 194
industrial organization, 170, 211
information, 7–9, 11–16, 28–30, 31–32, 37–43, 45–50, 62–63, 71–72, 79–80, 91, 126–129, 153, 157–159, 204, 238, 239–240
    as computation, 117–123, 136
    confusion over definition, 39–42
    dispersal of, 66, 157–158, 161, 196
    and epistemology, 8–9
    as an inductive index and/or the stochastic object of an epistemic logic, 108–117
    and intellectual property, 107
    mechanisms to reveal privately held, 135, 137, 171
    and the natural sciences, 31–32, 45–49, 109–112
    as public good, 106–107
    as supra human knowledge, 71
    taxonomy of formal approaches to, 102
    as thing or commodity, 78, 79, 103–107, 162
information revolution, 4, 20, 32
Internet, 32, 119
Isaac, Mark, 138–140

job shop problem, 95
Johnson, President Lyndon, 146
Journal of Economic Literature, 155
JP Morgan, 231

Kahn, Alfred, 138
Kashkari, Neel, 222
Katz, Michael, 214
Keynes, John Maynard, 78–79
knapsack problem, 213
Knight, Frank, 2–4, 6, 75, 78
knowledge, 1–3, 7, 34, 42–43, 55–57, 62–64, 67, 69–70, 77, 105–106, 109–110, 152, 237, 240, 242
    and information, 9, 20, 32, 37, 43, 48–49, 71, 103, 153, 239
    location of, 38, 152, 157–160, 179, 239
    shared, 115–116
    tacit, 67–68, 195
Koopmans, Tjalling, 74–75, 81–82, 95
Kramer, Gerald, 92
Krippner, Greta, 146
Krugman, Paul, 254
Kuhn, Thomas, 33–34

labor, 55, 78
Lancaster, Kelvin, 104
Lange, Oskar, 61, 67, 74, 76, 132, 137, 197
Lange-Lerner-Hayek controversy, 199
laws of supply and demand, 5, 42
Ledyard, John, 163, 202, 203

INDEX

Lehman Brothers, 222
Lerner, Abba, 76, 170, 197
Levine, Michael, 138–139, 256n35
Lewis, Alain, 92, 118
Lewis, Gregg, 74
liberalism, 3
logical positivism, 23

machine builders, 121, 215
Machlup, Fritz, 47–48, 104, 167
macroeconomics, 106, 152, 211
managerial economics, 172, 201
Mandelbrot, Benoit, 47
market design, 58, 118, 125–127, 130, 137, 138–139, 141–143, 147–150, 154–160, 170, 178–182, 184–188, 200, 204–205, 207–209, 217–219, 221, 238, 240
   and airport landing slots, 138–143, 185
   for particular markets, 125, 149–150, 241
   and the TARP, 223–232
Market Design Incorporated, 228–229, 265n32
market socialism, 76, 82, 88, 90, 134, 161, 170
markets, 7, 8, 54–55, 57–59, 67, 123, 125, 144–150, 155, 157–158, 179, 236
   abstract market, 241–242
   as communication devices, 137, 161–162, 164–165, 196
   engineered, 57–58
   epistemic capacities of, 7, 179
   as information processors, 7, 40, 54–56, 63, 64, 69–72, 127, 147, 160, 164, 179, 198–200, 225–226, 239–240
   in neoclassical theory, 25, 125, 144–145
   patenting, 219, 229
   as seat of supra-human knowledge, 70–72, 199
   as solutions to social problems, 57, 125, 207, 229
   as substitutes for regulation/bureaucracy, 140–141, 178, 184, 201, 217–218
Markets and Organizations report, 127, 138

Marschak, Jacob, 62, 67, 73–75, 79–80, 82–86, 89, 162–163, 172, 202–203, 250, 254
Marschak, Thomas, 163, 202–203
Marshall, Alfred, 36, 189–190
Marshallianism, 95, 189
Maskin, Eric, 131, 133, 193–194, 203, 204
Massachusetts Institute of Technology, 98–99, 106, 118, 163
McAfee, R. Preston, 214
McCabe, Kevin, 205
McCarthy, John, 84
MCI, 214, 216
McMillan, John, 241, 264
mechanism design, 87–89, 95–97, 132–137, 149–150, 156, 162, 163, 166, 168, 175–177, 184, 188, 193–194, 201–202, 255n22, 255n24, 261n7
mendacity, problem of, 171
Merrill Lynch, 231
message systems, 163–168
Milgrom, Paul, 172, 182, 190, 203, 214, 216–217, 219–220
Mises, Ludwig von, 60–62, 66, 75, 132, 134, 137
Mont Pelèrin Society, 7, 51–59, 61, 63–64, 147
Morgan Stanley, 231
mortgage backed securities, 146, 222, 224–225
Mosak, Jacob, 74
Mount, Kenneth, 96, 136, 163, 168, 196, 203
Myerson, Roger, 131–133, 194

NASA, 205
Nash equilibrium, 167, 176–177
Nash game theory, 40
national resident matching programs, 188
National Science Foundation, 96, 136, 201
natural selection, 69
negentropy, 48
neoclassical economics, 41, 5, 61, 63–64, 67, 73–123, 125, 147, 151, 169

INDEX

neoliberalism, 17, 27, 53–59, 61, 63, 99, 107, 127–128, 147, 160, 168, 236, 239–240, 242, 247n3
Neurath, Otto, 60
neuroeconomics, 17
Newell, Allen, 90
Noahpinion (blog), 235, 268n1
Nobel Prize (Bank of Sweden Prize), 51–52, 131, 161, 194
Noll, Roger, 139
Northwestern University, 124, 136, 172–173, 201, 214
NSF Conferences on Econometrics and Mathematical Economics, 25, 201
Nynex, 214

Office of Naval Research, 28, 172
operations research, 49, 76–77, 79, 108–109, 172–173, 183
organization theory, 89–91, 124, 128, 143
Osana, Hiroaki, 163

Pacific Bell, 214
Pardo-Guerra, Juan Pablo, 146
Pareto, Vilfredo, 26–27
Pareto optimality, 81, 120, 164, 166, 254n8
Paulson, Henry, 222, 224, 230, 231, 266
performativity theory, 148–149
physics, 24–25, 104, 109–110
Plott, Charles, 98, 138–141, 183–185
Polanyi, Michael, 68, 248
policymaking, 127, 208, 211–215
Polinomics, 140–141, 257
Poon, Martha, 146
positivism, 3
Posner, Richard, 58
postmodernism, 1, 240
Power Auctions LLC, 228–229, 265n32, 265n34
private values, 173–176, 180, 181, 190
privatization, 58, 127, 147
probability theory, 21, 23, 79
Project Revere, 49

psychology, 17, 18, 21, 22, 23, 26, 47–48, 67–68, 76, 90, 92–93, 108, 163
Purdue University, 95–96, 124, 136, 139, 201

Quinn, Sarah, 146
Quirk, James, 95

Radner, Roy, 92, 94, 126, 202–203
Raiffa, Howard, 171–172
RAND, 28, 74, 77, 79, 83, 90, 91, 98–99, 108, 163, 171–172, 251n49, 252n11
Rassenti, Stephen, 121, 142, 183–187, 205
rational choice model, 18–23, 26–29
   attempts to subsume under a computationalist model, 118
rationality, 16–24, 26–29, 31–32, 34–36, 40, 68, 86, 117, 128–129, 238
   bounded, 92–93, 167
   unconscious, 26, 68–69, 108, 248n7
Reagan, Ronald, 96, 127
reflexivity, 35, 105, 245n8, 252n2
Reiter, Stanley, 79, 80, 94–98, 119, 122, 126, 136, 162
   model of communication, 164–165, 168, 173, 196, 203
Riker, William, 139
Road to Serfdom (book), 75
Robbins, Lionel, 35
Romer, Paul, 53, 247n4
Roth, Alvin, 156, 183, 189, 191, 203, 219
Rothkopf, Michael, 173, 177
Royal Swedish Academy of Sciences, 131
Rumsfeld, Donald, 70, 249n11
Ryle, Gilbert, 68

Samuelson, Paul, 52, 63–64, 106
Sapienza, Paola, 3
Saposnik, Rubin, 95
Scarf, Herbert, 87
Schelling, Thomas, 41, 111
Schiller, Dan, 39
Schwartz, Nancy, 95
science studies, 148–149
securitization of mortgage debt, 146
semantics, 162–163

INDEX

Sen, Amartya, 20, 33
The Sensory Order (book), 67, 152
set packing problem, 186
Shannon, Claude, 46, 79, 104, 151, 166
Shannon information theory, 46–49, 89, 104–105, 107
  reconstruction by Walrasian school, 162, 166
Shell, 177
Simon, Herbert, 20, 63, 79, 80, 89–92, 117, 119, 152, 167
Sissoko, Carolyn, 146
smart markets, 188, 198–199, 213
Smith, Noah, 235, 268n1
Smith, Vernon, 95–97, 120–123
social choice theory, 21, 170
socialism, 55, 60, 61–62, 88, 124
Socialist Calculation Controversy, 60–65, 75
sociology, 21, 26, 148, 248n8
  Weberian, 62, 247, 253
Solow, Robert, 52, 63
Sonnenschein, Hugo, 5, 95
Sonnenschein/Mantel/Debreu theorems, 5
special investment vehicles, 146
spontaneous order, 71
Stanford University, 99, 171–172, 214
  Graduate School of Business, 171–172
State Street Bank, 231
statistics, 78–79, 102, 108, 111, 151
Stigler, George, 39–40, 52–53
Stiglitz, Joseph, 37, 41, 53, 65, 163
Sunder, Shyam, 122–123, 188
Swagel, Phillip, 222–223
Systems Research Lab, 91

TARP (Troubled Asset Relief Program), 221–232
team theory, 83–85
telecommunications, 103, 161–162, 207, 214. *See also* electromagnetic spectrums licenses
Telephone and Data Systems (TDS), 214
Texas, University of, 214
thermodynamics, 45
tort law, 59

toxic assets, 222, 225, 227–228, 231
truth, 2, 6, 7–8, 12–13, 34, 135–136, 179, 240, 242
Turing Machine, 117, 119, 152
types as models for information
  uncertainty, 114–116, 174, 177

UCLA, 172
uncertainty, 40–41, 47–48, 82, 86, 111, 113, 173–175
US Treasury, 223–224, 226, 228–232
"The Use of Knowledge in Society" (essay), 62, 137, 194
USSR, 124
utility functions, 19
utility theory, 23–25, 77–79, 92, 104, 111–113, 175

Veblen, Thorstein, 20, 93
Vickrey, William, 170–171, 173, 197
Vickrey-Clarke-Groves mechanism, 194
Vienna School, 60–61
von Neumann, John, 91, 98
von Neumann-Morgenstern expected utility, 23, 40, 111, 113
voting, 139–140

Wallis, Allen, 61
Walras, Leon, 145
Walrasian School, 4, 81, 87, 93–97, 99, 159, 161–169, 185, 195–196
Warsh, David, 247n4
Washington, University of, 49
Weber, Max, 62
Weber, Robert, 214
Wells Fargo, 231
Wiener, Norbert, 245
Williams, Arlington, 121, 185
Wilson, Robert, 171–173, 178, 197, 203, 214
Winner's Curse, 178, 228, 267n18

Yale University, 74, 84, 98, 99

zero-intelligence agents, 122
Zingales, Luigi, 3–4